ACCOUNTING *and* FINANCE *for the* NONFINANCIAL EXECUTIVE

An Integrated Resource Management Guide for the 21st Century

The St. Lucie Press
Library of Executive Excellence Series

ACCOUNTING *and* FINANCE *for the* NONFINANCIAL EXECUTIVE

An Integrated Resource Management Guide for the 21st Century

JAE K. SHIM, Ph.D.

Professor of Business Administration
California State University at Long Beach, California

St. Lucie Press
Boca Raton London New York Washington, D.C.

Library of Congress Cataloging-in-Publication Data

Shim, Jae K.
Accounting and finance for the nonfinancial executive : an integrated resource management guide for the 21st century / Jae K. Shim.
p. cm.— (The library for executive excellence)
Includes index.
ISBN 1-57444-287-2 (alk. paper)
1. Accounting. 2. Corporations—Finance. I. Title. II. Series.
HF5635.S552899 2000
657'.024'655—dc21 00-039041
CIP

Direct all inquiries to CRC Press LLC, 2000 N.W. Corporate Blvd., Boca Raton, Florida 33431.

Visit the CRC Press Web site at www.crcpress.com

St. Lucie Press is an imprint of CRC Press LLC

International Standard Book Number 1-57444-287-2
Library of Congress Card Number 00-039041
Printed in the United States of America 3 4 5 6 7 8 9 0
Printed on acid-free paper

Preface

This book is directed toward the businessperson who must have financial and accounting knowledge but has not had formal training in finance or accounting — perhaps a newly promoted middle manager or a marketing manager of a small company who must know some basic finance concepts. The entrepreneur or sole proprietor also needs this knowledge; he or she may have brilliant product ideas, but not the slightest idea about financing.

The goal of the book is to provide a working knowledge of the fundamentals of finance and accounting that can be applied, regardless of the firm size, in the real world. It gives nonfinancial managers the understanding they need to function effectively with their colleagues in finance.

We show you the strategies for evaluating investment decisions such as return on investment analysis. You will see what you need to know, what to ask, which tools are important, what to look for, what to do, how to do it, and what to watch out for. You will find the book useful and easy to read. Many practical examples, illustrations, guidelines, measures, rules of thumb, graphs, diagrams, and tables are provided to aid comprehension of the subject matter.

You cannot avoid financial information. Profitability statements, rates of return, budgets, variances, asset management, and project analyses, for example, are included in the nonfinancial manager's job.

The financial manager's prime functions are to plan for, obtain, and use funds to maximize the company's value. The financial concepts, techniques, and approaches enumerated here can also be used by any nonfinancial manager, irrespective of his or her primary duties.

This book is designed for nonfinancial executives in every functional area of responsibility in any type of industry. Whether you are in marketing, manufacturing, personnel, operations research, economics, law, behavioral sciences, computers, personal finance, taxes, or engineering, you must have a basic knowledge of finance. Because your results will be measured in dollars and cents, you must understand the importance of these numbers so as to optimize results in both the short and long terms.

Knowledge of the content of this book will enable you to take on additional managerial responsibilities. You will be better equipped to prepare, appraise, evaluate, and approve plans to accomplish departmental objectives. You will be able to back up your recommendations with carefully prepared financial support as well as state your particular measure of performance. By learning how to think in terms of finance and accounting, you can intelligently express your ideas, whether they are based on marketing, production, personnel, or other concepts.

You will learn how to appraise where you have been, where you are, and anywhere you are headed. Financial measures show past, current, and future performance. Criteria are presented to examine the performance of your division and product lines, and also formulate realistic profit goals.

Nonfinancial managers should have a grasp of financial topics, but need not be able to arrive at the mathematical answer (e.g., discounted rate of return problem). Nonfinancial managers mainly need to know enough to *ask* their financial colleagues what the discounted rate of return is for a variety of investment decisions. A decision can then be based on their answer.

You should have a basic understanding of financial information so as to evaluate the performance of your responsibility center. Are things getting better or worse? What are the possible reasons? Who is responsible? What can you do about it?

You need to know whether your business segment has adequate cash flow to meet requirements. Without adequate funds, your chances of growth are restricted.

You must know what your costs are in order to establish a suitable selling price. What sales are necessary for you to break even?

You may have to decide whether it is financially advantageous to accept an order at below the normal selling price. If you have idle facilities, a lower price may still result in profitability.

You need to be able to express your budgetary needs in order to obtain proper funding for your department. You may have to forecast future sales, cash flows, and costs to see if you will be operating effectively in the future.

You will spot areas of inefficiency or efficiency by comparing actual performance to standards through variance analysis. What are the reasons that sales targets differ from actual sales? Why are costs much higher than expected? The causes must be searched out so that corrective action may be taken.

You can undertake certain strategies to improve return on investment by enhancing profitability or using assets more efficiently. You have to understand that money is associated with a time value. Thus, you would prefer projects that generate higher cash flows in earlier years. You may also want to compute growth rates.

You are often faced with a choice of alternative investment opportunities. You may have to decide whether to buy machine A or machine B, whether to introduce a certain product line, or whether to expand.

In managing working capital, you have to get the most out of your cash, receivable, and inventory. How do you get cash faster and delay cash payments? Don't forget that you need liquid funds to meet ongoing expenditures. Should you extend credit to marginal customers? How much inventory should you order at one time? When should you order the inventory?

In financing the business, a decision has to be made whether short-term, intermediate-term, or long-term financing is suitable. The financing mix of the company in terms of equity of debt affects the cost of financing and influences the firm's risk position. What is the best financing source in a given situation?

Taxes are important in any business decision; the after-tax effect is what counts. Proper tax planning will make for wise decisions. Are you maximizing your allowable tax deductions?

Financial decisions are usually formulated on the basis of information generated by the accounting system of the firm. Proper interpretation of the data requires an understanding of the assumptions and rules underlying such systems, the convention adopted in recording information, and the limitation inherent in the information presented. To facilitate this understanding, an understanding of basic accounting concepts and conventions is helpful. You should be able to make an informed judgment on the financial position and operating performance of the entity. The balance sheet, the income statement, and the statement of cash flows are the primary documents analyzed to determine the company's financial condition. These financial statements are included in the annual report.

What has been the trend in profitability and return on investment? Will the business be able to pay its bills? How are the receivables and the inventory turning over? Various financial statement analysis tools are useful in evaluating the company's current and future financial conditions. These techniques include horizontal, vertical, and ratio analysis.

Keep this book handy for easy reference throughout your career; it will help you answer financial questions in all the areas mentioned here and in any other matter involving money.

Jae K. Shim

About the Author

Jae K. Shim is Professor of Accountancy and Finance at California State University, Long Beach. He received his M.B.A. and Ph.D. degrees from the University of California at Berkeley (Haas School of Business).

Dr. Shim is a coauthor of *Handbook of Financial Analysis, Forecasting, and Modeling, Encyclopedic Dictionary of Accounting and Finance, Barron's Accounting Handbook, Financial Accounting, Managerial Accounting, Financial Management, Strategic Business Forecasting, The Vest-Pocket CPA, The Vest-Pocket CFO*, and the best selling *Vest-Pocket MBA*. Dr. Shim has 45 other professional and college books to his credit.

Dr. Shim has also published numerous refereed articles in such journals as *Financial Management, Advances in Accounting, Corporate Controller, The CPA Journal, CMA Magazine, Management Accounting, Econometrica, Decision Sciences, Management Science, Long Range Planning, OMEGA, Journal of Operational Research Society, Journal of Business Forecasting,* and *Journal of Systems Management*. He was a recipient of the 1982 *Credit Research Foundation Outstanding Paper Award* for his article on cash budgeting.

Table of Contents

Part I Thinking Finance

Part II Critical Asset Management Issues

Part III Financial Decision Making for Managers

Part IV Obtaining Funds

Part V Dissecting Financial Statement Information

Part I

Thinking Finance

1 Financial Decision Making and Analysis

A company exists to increase the wealth of its owners. Management is concerned with determining which products and services are needed and putting them into the hands of its customers. Financial management deals with planning decisions to achieve the goal of maximizing the owners' wealth. Because finance is involved in every aspect of a company's operations, nonfinancial managers, like financial managers, cannot carry out their responsibilities without accounting and financial information.

In this chapter, you will learn about the nonfinancial manager's concern with finance, the scope and role of finance, the language of finance, the responsibilities of financial managers, the relationship between accounting and finance, and the financial and operating environment in which finance is situated.

1.1 THE NONFINANCIAL MANAGER'S CONCERN WITH FINANCE

You should have knowledge of finance and know how to apply it successfully in your particular departmental functions. This is true whether you are a manager in production, marketing, personnel, operations, or any other department. You should know what to look for, the right questions to ask, and where to get the answers. Financial knowledge aids in planning, problem solving, and decision making. Finance provides a road map in numbers and analysis so that you can optimally perform your duties. Further, you must have financial and accounting knowledge in order to understand the financial reports prepared by other segments of the organization. You must know what the numbers mean even if you do not have to determine them.

Nonfinancial managers spend a good portion of their time planning. They set objectives and plot efficient courses of action to obtain those objectives. There are many types of plans a nonfinancial manager might have to deal with: production plans, financial plans, marketing plans, personnel plans, and so on. Each of these plans is very different, and all require some kind of financial knowledge.

Finance provides a link that facilitates communication among different departments. For example, the budget communicates overall corporate goals to the department managers so they clearly know what is expected of them; it also provides guidelines for how each department may conduct its activities. Most importantly, you as a department manager must present a strong case to upper management to justify budgetary allowances. You are typically a participant providing input when the budget is prepared. You must identify any problems with the proposed budget so they are rectified before the budget is finalized. Even after the budget is implemented, you may suggest changes in subsequent budgetary formulations. Also, you must *intelligently discuss* the budget with other organizational members. If you do not adequately understand the budget or communicate requirements, your department may fail to achieve its goals.

You have to formulate and provide upper management with documented information to obtain approval for activities and projects (e.g., new product lines). Your request for resources will entail financial plans for the contemplated project. Here, a knowledge of forecasting and capital budgeting (selecting the most profitable of several alternative long-term projects) is required. You may be involved in a decision of whether to lease or buy an asset, such as equipment or an automobile. Thus, you must consider the feasibility of the purchase. You must evaluate and appraise monetary and manpower requests before submission. If you show signs of being ill prepared, you will give a negative impression that may result in the loss of resources.

In certain situations you may obtain financial information about competitors. You should be able to understand such data in order to make intelligent decisions.

Because many of your decisions have financial implications, you are continually interacting with financial managers. For instance, marketing decisions influence growth in sales and, as a result, there will be changes in plant and equipment requirements that dictate increased external funding. Thus, the marketing manager must have knowledge of the constraints of fund availability, inventory policies, and plant utilization. The purchasing manager must know whether sufficient funds exist to take advantage of volume discounts. The cost of raw materials is one of the most important manufacturing costs. The cost of alternative materials along with their quality must be known since cost affects selling price, and inferior materials may create production problems that eat into divisional profitability. Further, if materials are not delivered on time, customer orders may not be filled in a timely fashion, thus adversely affecting future sales. Advertising managers also make key decisions related to finance. They can justify costs associated with an advertising campaign by estimating its value. If customers want to buy your products, you have something of value that will pay off in future earnings.

Capital investment projects (property, plant, and equipment) are closely tied to plans for product development, marketing, and production. Thus, managers in these areas must be involved with planning and analyzing project performance. As one nonfinancial manager I interviewed who was working for an electronics company put it:

> My knowledge of accounting and finance helps me to report results, understand reports, control expenses, allocate resources, budget for proper staffing, and decide the direction of my department. There are thirty nonfinancial managers at my level within the company, and we work in a very competitive environment as the company only promotes from within. Therefore, I need every edge I can get in order to continue moving ahead, and my financial knowledge is a very important tool in my career development.

For these reasons, as well as a host of others, you need basic financial knowledge to successfully conduct daily activities.

1.2 WHAT ARE THE SCOPE AND ROLE OF FINANCE?

In this section, you will learn the language of finance as well as the what and why. You will see the responsibilities of financial managers, and the relationship between accounting and finance will be explained.

1.3 THE IMPORTANCE OF FINANCE

Finance provides discipline to all the components of the organization involved in decision making. Therefore, you need knowledge of it to perform effectively. A knowledge of finance terminology, concepts, techniques, and applications aids in the overall management of your departmental affairs.

For effective communication, you must be able not only to understand what financial people are saying, but also to express your ideas in their language. You can "open the door" to the finance department by having a better understanding of the finance function, thus leading to more productive working relationships with finance professionals.

If you master the finance vocabulary, you will be able to comprehend financial information (e.g., budgets), use that information effectively, and communicate clearly about the quantitative aspects of performance and results. You must clearly and thoughtfully express what you need to financial officers in order to perform effectively. To do so, you have to be familiar with the *basics* of accounting, taxes, economics, and other aspects of finance.

Finance uses *accounting information* to make decisions regarding the receipt and use of funds to meet corporate objectives. Accounting is generally broken down into two categories: financial accounting and managerial accounting. Financial accounting records the financial *history* of the business and involves the preparation of reports for use by external parties such as investors and creditors. Managerial accounting provides financial information useful in making better decisions regarding the future. Financial and managerial accounting are discussed later in this chapter. Chapters 19 to 21 cover financial accounting while Chapters 2 to 7 and 13 to 14 zero in on managerial accounting.

1.3.1 THE WHAT AND WHY OF FINANCE

Finance involves many interrelated areas such as obtaining funds, using funds, and monitoring performance. It enables you to look at current and prospective problems and find ways of solving them.

One important aspect of finance is the analysis of the return-risk tradeoff, which helps to determine if the expected return is sufficient to justify the risks taken. The greater the risk with any decision (e.g., new product line, new territory), the greater the return required. In managing your inventory of stock, for example, the less inventory (merchandise held for resale) you keep, the higher the expected return (since less cash is tied up), but also the greater the risk of running out of stock and thus losing sales and customer goodwill.

No matter who you are, you are involved with finance in one way or another. Financial knowledge is required of marketing managers, production personnel, business managers, investment planners, economists, public relations managers, operations research staff, lawyers, and tax experts, among others. For example, marketing managers have to know product pricing and variance analysis. Financial managers must know how to manage assets so as to optimize the rate of return. Production managers have to be familiar with budgeting and effective handling of productive assets. Personnel executives must know about planning. Public relations managers must know about the financial strengths of the business. Operations research staff has to know about the time value of money. Investment planners have to be familiar with the valuation of stocks, bonds, and other investments.

1.3.2 What Are Financial Managers Supposed to Do?

The financial manager plays an important role in the company's goals, policies, and financial success. The financial manager's responsibilities include the following:

- **Financial analysis and planning** — Determining the proper amount of funds to employ in the firm, that is, designating the size of the firm and its rate of growth.
- **Investment decisions** — Allocating funds to specific assets (things owned). The financial manager makes decisions regarding the mix and type of assets acquired, as well as modification or replacement of assets.
- **Financing and capital structure decisions** — Raising funds on favorable terms, that is, determining the nature of the company's liabilities (obligations). For instance, should funds be obtained from short-term or long-term sources?
- **Management of financial resources** — Managing cash, receivables, and inventory to accomplish higher returns without undue risk.

The financial manager affects stockholder wealth maximization by influencing:

1. Present and future earnings per share (EPS);
2. Timing and risk of earnings;
3. Dividend policy;
4. Manner of financing.

Table 1.1 presents the functions of the financial manager.

1.3.3 What Is the Relationship Between Accounting and Finance?

Accounting is a necessary input and subfunction to finance. The primary distinctions between accounting and finance relate to the treatment of funds and decision making.

If you are employed by a large firm, the financial responsibilities are probably carried out by the treasurer, controller, and financial vice president (chief financial officer). Figure 1.1 shows an organization chart of the finance structure within a

TABLE 1.1
Functions of the Financial Manager

A. Planning
- Long- and short-range financial and corporate planning
- Budgeting for operations and capital expenditures
- Evaluating performance
- Pricing policies and sales forecasting
- Analyzing economic factors
- Appraising acquisitions and divestment

B. Provision of Capital
- Short-term sources; cost and arrangements
- Long-term sources; cost and arrangements
- Internal generation

C. Administration of Funds
- Cash management
- Banking arrangements
- Receipt, custody, and disbursement of companies' securities and moneys
- Credit and collection management
- Managing pension moneys
- Investment portfolio management

D. Accounting and Control
- Establishing accounting policies
- Development and reporting of accounting data
- Cost accounting
- Internal auditing
- System and procedures
- Government reporting
- Report and interpretation of results of operations to management
- Comparison of performance with operating plans and standards

E. Protection of Assets
- Providing for insurance
- Establishing sound internal controls

F. Tax Administration
- Establishing tax policies
- Preparation of tax reports
- Tax planning

G. Investor Relations
- Maintaining liaison with the investment community
- Counseling with analyst-public financial information

H. Evaluation and Consulting
- Consultation with and advice to other corporate executives on company policies, operations, objectives, and their degree of effectiveness

I. Management Information Systems
- Development and use of computerized facilities
- Development and use of management information systems
- Development and use of systems and procedures

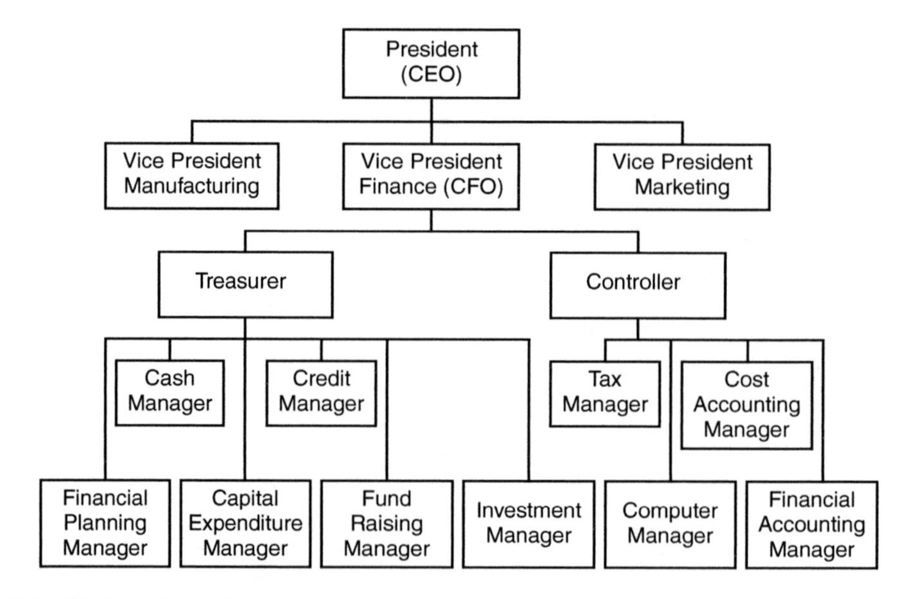

FIGURE 1.1 Financial activity organization.

company. Note that the controller and treasurer report to the vice president of finance. You should know the responsibilities of these financial officers within your own organization and how the function of each affects you.

The financial vice president is involved with financial policy making and planning. He or she has financial and managerial responsibilities, supervises all phases of financial activity, and serves as the financial adviser to the board of directors.

The effective, competent, and timely handling of controllership and treasurer functions will ensure corporate success. Table 1.2 lists the typical responsibilities of the treasurer and controller, but there is no universally accepted precise distinction among the two jobs. The functions may differ slightly between organizations because of personality and company policy, but typically controllers are concerned with *internal functions* whereas treasurers are responsible for *external functions*.

Management is involved with finance primarily in two ways. First, there is the record keeping, tracking, and controlling of the financial effects of prior and present operations, as well as obtaining funds to satisfy current and future requirements; this function is of internal nature. The external function involves outside entities.

The internal matters of concern to the controller include financial and cost accounting, taxes, control, and audit functions. The controller is primarily involved in collecting and presenting financial information. He or she typically looks at what *has* happened instead of what should or will happen. The controller prepares the annual report and Securities and Exchange Commission (SEC) filings as well as tax returns. The SEC filings include Form 10-K, Form 10-Q, and Form 8-K. The primary function of the controller is ensuring that funds are used efficiently.

The control features of the finance function are referred to as *managerial accounting*. Managerial accounting is the preparation of reports used by management for internal decision making, including budgeting, costing, pricing, capital

TABLE 1.2
Responsibilities of Controller and Treasurer

Controller	Treasurer
Accounting	Obtaining financing
Reporting of financial information	Banking relationship
Custody of records	Investment of funds
Interpretation of financial data	Investor relations
Budgeting	Cash management
Controlling operations	Insuring assets
Appraisal of results and making recommendations	Fostering relationship with creditors and investors
Preparation of taxes	Credit appraisal and collecting funds
Managing assets	
Internal auditing	Deciding on the financing mix
Protection of assets	Dividend disbursement
Reporting to the government	Pension management
Payroll	

budgeting, performance evaluation, break-even analysis (sales necessary to cover costs), transfer pricing (pricing of goods or services transferred between departments), and rate of return analysis. Managerial accounting relies heavily on historical information generated in *the financial accounting* function, but managerial accounting differs from financial accounting in that it is future-oriented (making decisions that ensure future performance).

Managerial accounting information is vital to the nonfinancial person. For example, the break-even point is useful to marketing managers in deciding whether to introduce a product line. Variance analysis is used to compare actual revenue and costs to standard revenue and costs for performance evaluation so that inefficiencies can be identified and collective action taken. Budgets provide manufacturing guidelines to production managers.

Many controllers are involved with management information systems that analyze prior, current, and emerging patterns. The controller function also involves reporting to top management and analyzing the financial implications of decisions.

The treasurer's responsibility is mostly custodial in obtaining and managing the company's capital. Unlike the controller, the treasurer is involved in external activities primarily involving financing matters. He or she deals with creditors (e.g., bank loan officers), stockholders, investors, underwriters for equity (stock) and bond issuances, and governmental regulatory bodies such as the SEC. The treasurer is responsible for managing corporate assets (e.g., accounts receivables inventory) and debt, planning the finances, planning capital expenditures, obtaining funds, formulating credit policy, and managing the investment portfolio.

The treasurer concentrates on keeping the company afloat by obtaining cash to meet obligations and to buy the assets needed to achieve corporate objectives. Whereas the accountant concentrates on profitability, the treasurer emphasizes the sources and uses of cash flow. Even a company that has been profitable may have a significant negative cash flow. For example, there may exist substantial long-term receivables (debts owed to the company but not yet paid). In fact, without sufficient cash flow, a company may fail. By concentrating on cash flow, the financial manager should prevent bankruptcy and accomplish corporate goals. The financial manager appraises the financial statements, formulates additional data, and makes decisions based on the analysis.

1.4 FINANCIAL AND OPERATING ENVIRONMENT

You operate in the financial environment and are indirectly affected by it. In this section we will discuss financial institutions and markets, financial vs. real assets, and the alternative forms of business organizations.

1.4.1 What Should You Know About Financial Institutions and Markets?

A healthy economy depends heavily on efficient transfer of funds from savers to individuals, businesses, and governments who need capital. Most transfers occur through *specialized financial institutions*, which serve as intermediaries between suppliers and users of funds.

In the financial *markets*, companies demanding funds are brought together with those having surplus funds. Financial markets provide a mechanism through which the financial manager obtains funds from a wide range of sources, including financial institutions. The financial markets are composed of money markets and capital markets. Figure 1.2 depicts the general flow of funds among financial institutions and markets.

Money markets are the markets for short-term (less than one year) debt securities. Examples of money market securities include U.S. Treasury bills, commercial paper, and negotiable certificates of deposit issued by government, business, and financial institutions.

Capital markets are the markets for long-term debt and corporate stocks. The New York Stock Exchange, which handles the stocks of many of the larger corporations, is an example of a major capital market. The American Stock Exchange and the regional stock exchanges are still other examples. In addition, securities are traded through the thousands of brokers and dealers on the *over-the-counter (OTC) market*, a term used to denote all buying and selling activities in securities that do not occur on an organized stock exchange.

1.4.2 Financial Assets vs. Real Assets

The two basic types of investments are financial assets and real assets. Your *financial assets* comprise intangible investments (things you cannot touch). They represent your equity ownership of a company, or they provide evidence that someone owes you a debt, or they show your right to buy or sell your ownership interest at a

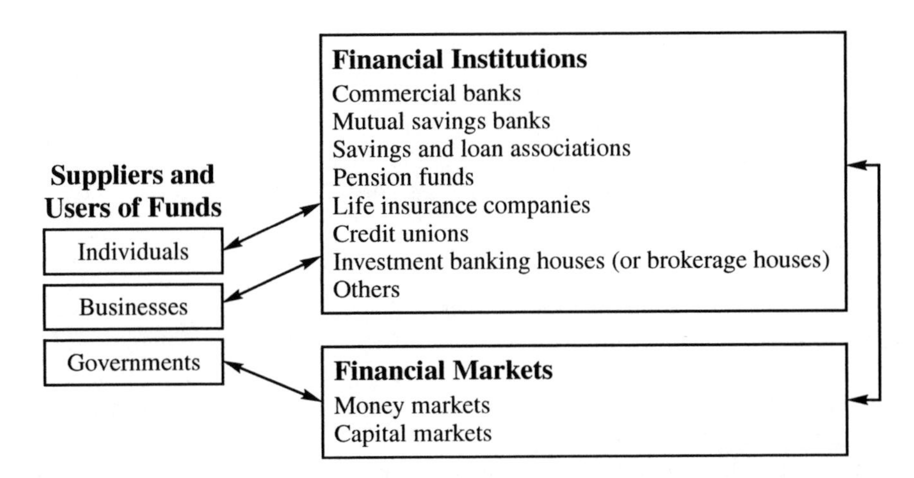

FIGURE 1.2 General flow of funds among financial institutions and financial markets.

subsequent date. Financial assets include common stock, options and warrants to buy stock at a later date, money market certificates, savings accounts, Treasury bills, commercial paper (unsecured short-term debt), bonds, preferred stock, and financial futures (contracts to buy financial instruments at a later date). *Real assets* are investments you can put your hands on. Sometimes referred to as real property, they include real estate, machinery and equipment, precious and common metals, and oil.

1.4.3 Basic Forms of Business Organizations

The basic types of businesses are sole proprietorship, partnership, and corporation. Of the three, corporations are usually of the largest size (in terms of sales, total assets, or number of employees), whereas partnerships and proprietorships emphasize entrepreneurship to a greater degree.

1.4.3.1 Sole Proprietorship

A sole proprietorship is owned by one individual. Sole proprietorships are the most numerous of the three types of organizations. The typical sole proprietorship is a small business; usually, only the proprietor and a few employees work in it. Funds are raised from personal resources or through borrowings. The sole proprietor makes all the decisions. Sole proprietorships are common in the retail, wholesale, and service sectors.

The advantages of a sole proprietorship are:

- No formal charter is required.
- Organizational costs are minimal.
- Profits and control are not shared with others.
- The income of the business is taxed as personal income.
- Confidentiality is maintained.

The disadvantages are:

- The ability to raise large sums of capital is limited.
- Unlimited liability exists for the owner.
- The life of the business is limited to the life of the owner.
- The sole proprietor must be a "jack-of-all-trades."

1.4.3.2 Partnership

A partnership is similar to the sole proprietorship except that the business has more than one owner. Partnerships are often formed to bring together different skills or talents, or to obtain the necessary capital. Although partnerships are generally larger than sole proprietorships, they are not typically large businesses. Partnerships are common in finance, real estate, insurance, public accounting, brokerage, and law.

The partnership contract (articles of partnership) spells out the rights of each partner concerning such matters as profit distribution and fund withdrawal. Partnership property is *jointly* owned. Each partner's interest in the property is based on his or her proportionate capital balance. Profits and losses are divided in accordance with the partnership agreement. If nothing about distribution is stated, they are distributed equally.

Each partner acts as an *agent* for the others. The partnership (and thus each individual partner) is legally responsible for the acts of any partner. However, the partnership is not bound by acts committed beyond the scope of the partnership. Forming a partnership creates these advantages:

- Partnerships can be easily established, with minimal organizational effort.
- Partnerships are free from special governmental regulation, at least compared to corporations.
- Income of the partnership is taxed as personal income to the partners.
- More funds are typically obtained than by a sole proprietorship.
- Better credit standing results from the availability of partners' personal assets to meet creditor claims.
- Partnerships attract good employees because of potential partnership opportunities.

Its disadvantages are as follows:

- It carries unlimited liability for the partners; each member is held *personally* liable for all partnership debts.
- It dissolves upon the withdrawal or death of any partner.
- Because it cannot sell stock, its ability to raise significant capital is limited, which may restrict growth.

1.4.3.3 Corporation

A corporation is a legal entity existing apart from its owners (stockholders). Ownership is evidenced by possession of shares of stock. The corporate form is not the

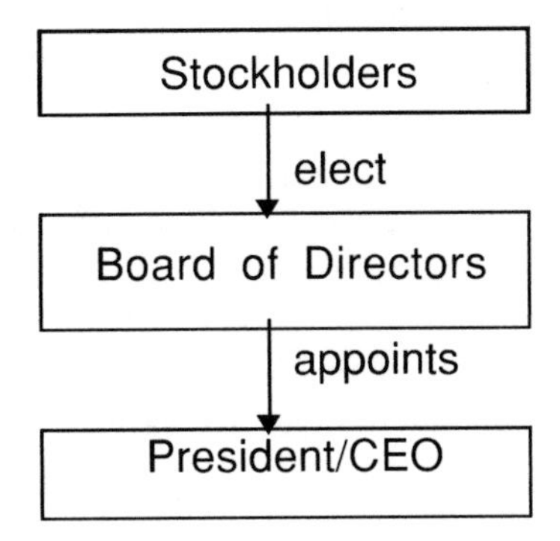

FIGURE 1.3 Corporate structure.

most numerous type of business, but it is the most important in terms of total sales, assets, profits, and contribution to national income. The corporate form is implicitly assumed throughout this book. Corporations are governed by a distinct set of state or federal laws and come in two forms: a *state C Corporation* or *federal Subchapter S Corporation*.

The advantages of a C corporation are:

a. Unlimited life.
b. Limited liability for its owners, as long as no personal guarantee on a business-related obligation such as a bank loan or lease exists.
c. Ease of transfer of ownership through transfer of stock.
d. Ability to raise large sums of capital.

Its disadvantages are:

a. Difficult and costly to establish, as a formal charter is required.
b. Subject to double taxation on its earnings and dividends paid to stockholders.
c. Bankruptcy, even at the corporate level, does not discharge tax obligations.

1.4.3.3.1 Subchapter S Corporation

A Subchapter S Corporation is a form of corporation whose stockholders are taxed as partners. To qualify as an S Corporation, the following is necessary:

a. A corporation cannot have more than 35 shareholders.
b. It cannot have any nonresident foreigners as shareholders.
c. It cannot have more than one class of stock.
d. It must properly elect Subchapter S status.

The S Corporation can distribute its income directly to shareholders and avoid the corporate income tax while enjoying the other advantages of the corporate form. *Note*: Not all states recognize Subchapter S Corporations.

The general structure of a corporation is depicted in Figure 1.3.

1.4.3.3.2 Limited Liability Company

Limited Liability Companies (LLCs) are a relatively recent development. LLCs are typically not permitted to carry on certain service businesses (e.g., law, medicine, and accounting). An LLC provides limited personal liability, as a corporation does. Owners, who are called members, can be other corporations. The members run the company unless they hire an outside management group. The LLC can choose whether to be taxed as a regular corporation or pass through to members. Profits and losses can be split among members any way they choose. Most states permit the establishment of LLCs. *Note*: Rules governing LLCs vary by state.

1.5 CONCLUSION

This chapter discussed the functions of finance, the environment in which finance operates, and how the nonfinancial manager fits in a typical company's structure.

The financial functions of the business impact nonfinancial activities in such areas as record keeping, performance evaluation, variance analysis, and the acquisition and utilization of resources. The nonfinancial manager must comprehend the goals, procedures, techniques, yardsticks, and functions of finance to optimally perform his or her duties. Ignorance of finance will not only lead to incorrect analysis and decisions but will also prevent you from moving within the organization.

An important reason why you need the financial and accounting knowledge edge is that without a good understanding of these disciplines you do not have the tools needed for effective management decision making. You would have to rely totally on the financial manager, whose recommendations you may not be able to totally understand or, if necessary, dispute. A successful operation blends sales, marketing, promotion, advertising, and finance with some degree of goal congruence. Decisions that make sense in terms of marketing and sales must also make financial sense. Without some financial background, you cannot contribute sound input to the decision process.

2 What Can You Do About Your Departmental Costs?

Cost is an expenditure incurred to obtain revenue. In establishing a price for your product or service, you must know total costs and costs per unit. You must also be familiar with the various cost concepts that are useful for income determination, short-term and long-term decision making, and planning, evaluation, and control. Different types of costs are used for different purposes, and proper costing will ensure the appropriate use of and accountability for your department's resources.

2.1 IMPORTANCE OF COST DATA

Obtaining and understanding cost information is essential to your business success. First, your costs determine your selling price; if your costs exceed your selling price, you will incur a loss. All costs applicable to a product or service must be considered (including manufacturing, selling, and other expenses) when determining a selling price that accounts for expected inflationary price increases. For example, if inflation is expected to be 6% next year, the selling price should similarly be increased by 6%.

Cost information also assists in determining: (1) the minimum order to be accepted; (2) the profitability of a particular product, territory, or customer; and (3) the method of servicing particular types of accounts (e.g., through jobbers, by telephone, or by mail order). In addition, cost information allows the purchasing manager to evaluate which suppliers are the least costly to use (i.e., the total cost associated with buying their merchandise, including any transportation charges). Cost information is also useful in planning and budgetary decisions. Your budgeted costs must be sufficient to meet your needs.

2.2 TYPES OF COSTS

Costs are classified according to function, ease of traceability, timing of charges against revenue, behavior, averaging, controllability, and other important cost concepts.

2.2.1 COSTS BY FUNCTION

Manufacturing costs include any costs to produce a product consisting of direct material, direct labor, and factory overhead.

Direct material becomes an integral part of the finished product (e.g., steel used to make an automobile).

Direct labor is labor directly involved in making the product (e.g., the wages of assembly workers on an assembly line).

Factory overhead includes all costs of manufacturing except direct material and direct labor (e.g., depreciation, rent, taxes, insurance, and fringe benefits).

Nonmanufacturing costs (operating expenses) are expenses related to operating the business rather than producing the product. These costs include sales, general, and administrative expenses.

Selling expenses include the cost of obtaining the sales commission, salespersons' salaries, or distributing the product to the customer (e.g., delivery charges). Selling costs may be analyzed for reasonableness by product, territory, customer class, distribution outlet, and method of sale. Marketing costs should be evaluated based on the success of distribution methods (e.g., direct selling to retailers and wholesalers vs. mail order sales).

General and administrative expenses include the costs incurred for administrative activities (e.g., executive salaries and legal expenses).

2.2.2 Costs by Ease of Traceability

Direct costs are directly traceable to a particular object of costing such as a department, product, job, or territory (e.g., depreciation on machinery in a department and advertising geared to a particular sales territory).

Indirect (common) costs are more difficult to trace to a specific costing object because they are shared by different departments, products, jobs, or territories. Therefore, such costs are allocated on a rational basis (e.g.. rent is allocated to a department based on square footage). A cost may be direct in one area and indirect in another. For example, in analyzing salespeople, traveling and entertainment expenses are direct; however, in an analysis by product, these expenses are indirect.

2.2.3 Costs by Timing of Charges Against Revenue

Period costs are related to time rather than to producing the product (e.g., advertising costs, sales commissions, and administrative salaries). They are charged against revenue in full in the year incurred.

Product costs are related to manufacturing a product (e.g., material and labor costs). They are charged to inventory first and then to cost of sales when sales are made.

2.2.4 Costs by Behavior

Fixed costs include the costs that remain constant regardless of activity (e.g., rent, property taxes, insurance). As sales increase, fixed costs do not increase; therefore, profits can increase rapidly during good times. However, during bad times fixed costs do not decline, which causes profits to fall rapidly.

Variable costs include the costs that vary directly with changes in activity (e.g., direct material, direct labor). Thus, a 20% increase in variable cost accompanies a 20% increase in sales.

Semivariable (mixed) costs include the costs that are part fixed and part variable. A semivariable cost varies with changes in volume but, unlike a variable cost, does not vary in direct proportion. Such examples include telephone bills, electricity bills, and car rentals charged as fixed rental fees plus variable mileage fees. The breakdown of costs into their variable and fixed components is important in many areas, including flexible budgeting, break-even analysis, and short-term decision making.

2.2.5 Costs by Averaging

Average costs are the total costs divided by total units. For example, if total cost is $10,000 for the production of 1,000 units, the average cost is $10 per unit.

2.2.6 Costs by Controllability

Controllable costs are those costs over which a manager has direct control (e.g., advertising if it is at the manager's discretion).

Uncontrollable costs are those costs over which a manager has no control (e.g., property taxes).

2.3 OTHER IMPORTANT COST CONCEPTS USEFUL FOR PLANNING, CONTROL, AND DECISION MAKING

Standard costs are the carefully predetermined production or operating costs that serve as target costs. Standard costs are compared with actual costs in order to measure the performance of a given department.

Joint costs are incurred for the benefit of the entire company (e.g., legal fees).

Incremental (differential) costs are determined by calculating the difference in costs between two or more alternatives. For example, if the direct labor costs for products A and B are $10,000 and $15,000, respectively, the incremental cost is $5,000.

Sunk costs are those costs of resources that have already been incurred, and, therefore, will not change regardless of which alternative is chosen. They represent past or historical costs. For example, a $50,000 machine paid for three years ago now has a book value of $20,000. This $20,000 book value is a sunk cost that does not affect a future decision.

Relevant costs include expected costs that will differ between alternatives. The incremental cost is relevant to a decision, but the sunk cost is irrelevant.

Opportunity costs include the net revenue foregone by rejecting an alternative. For example, if you have the choice of using your department's capacity to produce an extra 10,000 units or renting it out for $20,000, the opportunity cost of using the capacity is $20,000.

TABLE 2.1
Cost Behavior

	Unit Cost	Total Cost
Fixed	Up/down to volume	Constant
Variable	Constant	Up/down to volume
Semivariable	Up/down to volume	Up/down to volume

TABLE 2.2
Cost Behavior for Rent

Volume	Rent	Unit Cost
100,000	$100,000	$1.00
150,000	100,000	.67
200,000	100,000	.50

Discretionary costs are those costs that can be discontinued without affecting the accomplishment of essential managerial objectives in the short term (e.g., bonuses).

2.4 HOW DO YOUR COSTS BEHAVE?

2.4.1 Costs by Behavior

From a planning and control standpoint, perhaps the most important way to classify costs is by how they behave in accordance with changes in volume or some measure of activity. Assuming idle capacity (not using all your department's capacity), the cost behavior relationships of fixed cost, variable cost, and semivariable cost are shown in Table 2.1. Table 2.2 illustrates the cost behavior for a fixed cost, such as rent, and Figure 2.1 shows the behavior pattern for a fixed cost.

Example 2.1 — Your company is operating at idle capacity. Current production is 100,000 units. Total fixed cost is $100,000, and variable cost per unit is $3. If production increases to 110,000 units, the following results:

1. Total fixed cost is still $100,000.
2. Fixed cost per unit is $.91 ($100,000/110,000 units)
3. Total variable cost is $330,000.
4. Variable cost per unit is $3.

Table 2.3 illustrates the cost behavior for a variable cost, such as commissions, and Figure 2.2 shows the behavior pattern for a variable cost.

You can estimate the total cost of an item, such as a product line, by combining the variable cost and fixed cost.

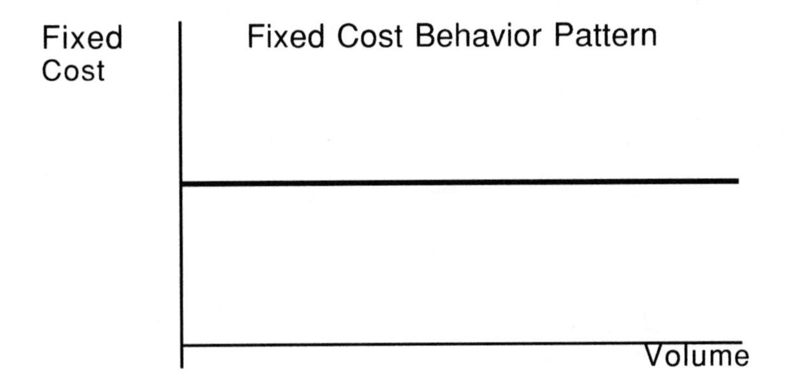

FIGURE 2.1 Fixed cost behavior pattern.

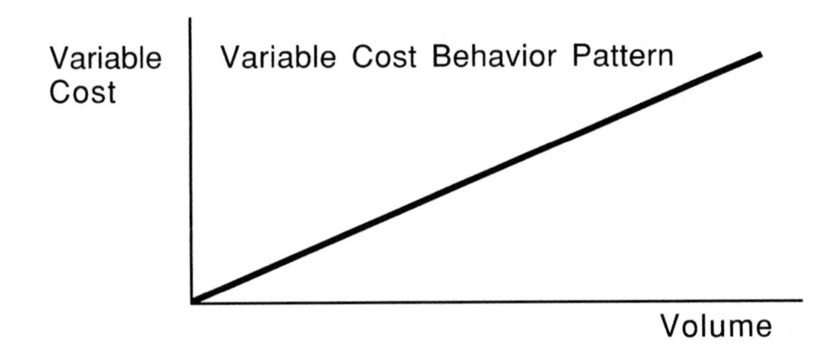

FIGURE 2.2 Variable cost behavior pattern.

TABLE 2.3
Cost Behavior for Commissions

Volume	Commissions	Unit Cost
100,000	$10,000	$.10
150,000	15,000	.10
200,000	20,000	.10

Example 2.2 — There are 100 estimated units for product line X. The fixed cost is $600, and the variable cost is $2.25 per unit. The total cost is:

Fixed cost	$ 600	
Variable	+ 225	(100 × $2.25)
Total cost	$ 825	

2.5 SEGREGATING FIXED COST AND VARIABLE COST

If you know the total cost you can determine the fixed and variable costs using the high-low method. The variable cost per unit may be computed by comparing the change in expense between high and low levels with the change in volume.

Example 2.3 — Your company reached its sales volume high in May. The lowest volume occurred in February.

	Total Cost	Unit Volume
High point (May)	$4,000	32,000
Minus low point (February)	2,800	20,000
Change due to volume	$1,200	12,000

$$\text{Change in cost per unit} = \frac{\$1,200}{\$12,000} = \$.10$$

Thus, the variable cost per unit is $.10. You can now determine the fixed portion of the expense using either the lowest or highest volume level. For example, using the high point:

$$\text{Total Fixed Cost} = \text{Total Cost} - \text{Total Variable Cost}$$

$$\text{Total Fixed Cost} = \$4,000 - (\$.10 \times 32,000) = \$800$$

2.6 COST ALLOCATION

Cost allocation assigns a common cost to two or more departments. The costs are allocated in proportion to each department's responsibility for their incurrence. Possible allocation bases include units produced, direct labor cost, direct labor hours, machine hours, and number of employees. Criteria in selecting an allocation base include cost benefit, ease of use, and industry standards. Cost allocation enhances control, aids efficiency evaluation, and promotes sound decision making. Accurate cost figures are necessary for product costing and pricing.

2.7 COST ANALYSIS

Cost analysis allows management to move carefully and accurately to control costs in the following ways:

- Compensation and expenses for salespeople should be analyzed and compared to budget, salary structure, and industry standards.
- Cost estimates may be made for alternative methods of selling products (e.g., evaluation can be performed related to the distribution of samples and the effect on costs and sales trends).
- Cost data of potential sales by product or territory may aid in the assignment of sales people.

- Cost information for advertising programs helps in making decisions for future media communications. Special cost structure may be formulated for market test cases to examine cost effectiveness.
- An analysis of entertainment expenses should be made by customer, salesperson, or territory, in order to determine whether expenses are in line with the revenue obtained and whether the cost per dollar of net sales and the cost per customer are reasonable.
- The following costs should also be analyzed for control purposes: cost per order received, cost per customer account, cost per item handled, cost per shipment, and cost per order filled.
- The cost per month and cost per mile for auto expenses should be considered for reasonableness.
- Material costs should be evaluated for changes. Costs per unit may drop as a result of quantity discounts and changes in the suppliers' freight charges and terms, or they may change due to substitution of different materials, changing suppliers, and difference in the quality of material.
- Direct labor costs should also be evaluated for changes. Costs per unit may decline due to increased worker experience in performing the task. Also, new material waste will decline as the operation attains increased maturity.

2.8 WHAT YOU CAN LEARN FROM THE JAPANESE

Japanese manufacturing companies place high standards on quality, timely delivery, and low-cost production. To lower costs they reduce the number of parts and use standard parts across product lines with a variety of products. Japanese companies also recognize that the best area to support low-cost production is usually in a product's design stage. They design and build products and sell them at prices that will ensure market success. Note that such selling prices may be lower than that supported by current manufacturing costs. In an attempt to improve machine efficiency, Japanese companies practice preventative and corrective maintenance, instead of breakdown maintenance.

2.9 CONCLUSION

Cost information is imperative in the decision-making process and is necessary for operational and control purposes as well. The project manager should have a basic knowledge of cost control so as to keep track of the expenses of his or her product. Such knowledge would be useful in analysis of payment requests to vendors, equipment purchasing and rentals, lump-sum vs. cost plus contracts, analysis of labor cost, preparation of the progress report, and evaluation of project status.

3 How You Can Use Contribution Margin Analysis

Contribution margin analysis is another tool managers use for decision making. In the contribution margin approach, expenses are categorized as either fixed or variable. The variable costs are deducted from sales to obtain the contribution margin. Fixed costs are then subtracted from contribution margin to obtain net income. This information helps the manager to: (1) decide whether to drop or push a product line; (2) evaluate alternatives arising from production, special advertising, and so on; (3) decide on pricing strategy and products or services to emphasize; and (4) appraise performance. For instance, this procedure would be useful to formulate a bid price on a contract, and to decide whether to accept an order even if it is below the normal selling price.

The format of the contribution margin income statement follows:

Sales	
Less:	Variable cost of sales
	Variable selling and administrative expenses
Contribution margin	
Less: Fixed cost	
Net income	

Example 3.1 — If the selling price is $10 per unit and the variable cost is $8 per unit, a contribution margin of $2 per unit is earned. The contribution margin ratio (contribution margin/sales) is 20% ($2/$10).

Example 3.2 — You sell 40,000 units of a product at $20 per unit. The variable cost per unit is $5, and the fixed cost is $250,000; the contribution margin income statement is as follows:

Sales (40,000 × $20)	$ 800,000
Less: Variable cost (40,000 × $5)	200,000
Contribution margin (40,000 × $15)	$ 600,000
Less: Fixed Cost	250,000
Net Income	$ 350,000

Example 3.3 — The following data applies to your department: sales $50,000, variable cost $45,000, and fixed cost $3,000. If there is an expected 10% increase in sales, the expected departmental income will be as follows:

Sales ($50,000 × 1.10)	$ 55,000
Less: Variable cost ($45,000 × 1.10)	49,500
Contribution margin	$ 5,500
Less: Fixed cost	3,000
Departmental income	$ 2,500

3.1 SHOULD YOU ACCEPT A SPECIAL ORDER?

When idle capacity exists, an order should be accepted at below the normal selling price, provided a contribution margin is earned, because fixed cost will not change.

Example 3.4 — You currently sell 8,000 units at $30 per unit. The variable cost per unit is $15, and the fixed cost is $60,000 (fixed cost per unit is $60,000/8,000, or $7.50). Idle capacity exists. A potential customer is willing to purchase 500 units at $21 per unit. The contribution margin income statement for this special order would be as follows:

Sales (500 × $21)	$ 10,500
Less: Variable cost (500 × $15)	7,500
Contribution Margin	$ 3,000
Less: Fixed cost	0
Net Income	$ 3,000

You should accept this order because it increases your profitability. If idle capacity exists, the acceptance of an additional order does not increase fixed cost. However, even if fixed costs were to increase, say, by $1,200 to buy a special tool for this job, it still would be financially attractive to accept this order because you would still realize a profit of $1,800 ($3,000 - $1,200).

Example 3.5 — Financial data for your department follows:

Selling price	$ 15.00
Direct material	$ 2.00
Direct labor	$ 1.90
Variable overhead	$ .50
Fixed overhead ($100,000/$20,000) units	$ 5.00

Selling and administrative expenses are fixed except for sales commissions, which are 14% of the selling price. Idle capacity exists. You receive an additional order for 1,000 units from a potential customer willing to pay $9 per unit. The contribution margin income statement for this special order would be as follows:

Sales (1,000 × $9)	$ 9,000
Less: Variable manufacturing cost (1,000 × $4.40)*	4,400
Manufacturing contribution margin	$ 4,600
Less: Variable selling and administrative expenses (14% × $9,000)	1,260
Contribution margin	$ 3,340
Less: Fixed cost	0
Net income	$ 3,340

* Variable manufacturing cost = $2.00 + $1.90 + $.50 = $4.40

Even though the offered selling price of $9 is much less than the current selling price of $15, the order should be accepted.

Example 3.6 — You want a markup of 40% over cost on a product. You can determine your selling price by using the following data:

Direct material	$ 5,000
Direct labor	12,000
Overhead	4,000
Total cost	$ 21,000
Markup on cost (40%)	8,400
Selling price	$ 29,400

Total direct labor for the year is $1,800,000. Total overhead for the year is 30% of direct labor. The overhead is 25% fixed and 75% variable. A customer offers to buy the item for $23,000. There is idle capacity. The following contribution margin income statement will show you whether you should accept the special offer.

Selling price		$ 23,000
Less: Variable costs		
Direct material	$ 5,000	
Direct labor	12,000	
Variable overhead ($12,000 × 22.5%)*	2,700	
	19,700	19,700
Contribution margin		$ 3,300
Less: Fixed cost		0
Net income		$ 3,300

* Total overhead 0.30 × $1,800,000 = $540,000; variable overhead = 22.5% of direct labor, computed as follows:

$$\frac{\text{Variable overhead}}{(\text{\% of direct labor})} = \frac{0.75 \times \$540,000}{\$1,800,000} = \frac{\$405,000}{\$1,800,000} = 22.5\%$$

In this case, you should accept the incremental order since it will be profitable. However, if you are not the one who decides whether to accept or reject special orders, you still need to understand why your company asks you to sell an item at a lower price.

3.2 HOW DO YOU DETERMINE A BID PRICE?

Pricing policies using contribution margin analysis may be helpful in contract negotiations. Often such business is sought during the slack season, when it may be financially beneficial to bid on extra business at a competitive price that covers all variable costs and makes some contributions to fixed costs plus profits. A knowledge of your variable and fixed costs is necessary to make an accurate bid price determination.

Example 3.7 — You receive an order for 10,000 units and want to know the minimum bid price that will result in a $20,000 increase in profits. The current income statement is as follows:

Sales (50,000 units × $25)		$1,250,000
Less: Cost of sales		
Direct material	$120,000	
Direct labor	200,000	
Variable overhead ($200,000 × 0.30)	60,000	
Fixed overhead	100,000	
	480,000	480,000
Gross margin		$770,000
Less: Selling and administrative expenses		
Variable (includes freight costs of $0.40 per unit)	$ 60,000	
Fixed cost	30,000	
	90,000	90,000
Net income		$ 680,000

Cost patterns for the incremental order are the same except that:

- Freight costs will be borne by the customer:
- Special tools costing $8,000 will be required and will not be reused again.
- Direct labor time for each unit under the order will be 20% longer.

Preliminary computations indicate the following per-unit costs:

Direct material ($120,000/50,000)	$2.40
Direct labor ($200,000/50,000)	4.00
Variable selling and administrative expense ($60,000/50,000)	1.20

A forecasted income statement, like the one on Table 3.1, provides more information on how to determine a bid price.

The contract price for the 10,000 units should be $122,400 ($1,372,400 – $1,250,000), or $12.24 per unit ($122,400/10,000). The contract price per unit of $12.24 is below the $25 current selling price per unit. Keep in mind that total fixed cost is the same except for the $8,000 expenditure on the special tool.

3.3 DETERMINING PROFIT FROM YEAR TO YEAR

Contribution margin analysis also aids in determining how to obtain the same profit as the previous year even with decreased sales volume.

Example 3.8 — In 2000, sales volume was 200,000 units, the selling price was $25, the variable cost per unit was $15, and the fixed cost was $500,000. In 2001, sales volume is expected to total only 150,000 units. As a result, fixed costs have been slashed by $80,000. In 2001, 40,000 units have already been sold. You wish to compute the contribution margin that must be earned on the remaining units for 2001 in order to make the same net income as in 2000.

TABLE 3.1
A Forecasted Income Statement

	Current	Projected	
Units	50,000	60,000	
Sales	$1,250,000	$1,372,400[a]	Computed last
Cost of sales			
Direct material	$120,000	$144,000	($2.40 × 60,000)
Direct labor	200,000	248,000	($200,000 + [10,000 × $4.80[b]])
Variable overhead	60,000	74,400	($248,000 × 30%)
Fixed overhead	100,000	108,000	
Total	$480,000	$574,400	
Selling and administration costs			
Variable	$60,000	$68,000	($60,000 + [10,000 × $0.80[c]])
Fixed	30,000	30,000	
Total	$90,000	$98,000	
Net income	$680,000	$700,000[d]	

[a]Net income + selling and administrative expense + cost of sales = sales: $700,000 + $98,000 + $574,400 = $1,372,400
[b]$4 × 1.20 = $4.80
[c]$1.20 – $0.40 = $0.80
[d]$680,000 + $20,000 = $700,000

Net income (P) computation for 2000:

$$\text{Sales} = \text{Fixed Cost} + \text{Variable Cost} + P$$

$$\$25 \times 200,000 = \$500,000 + (\$15 \times 200,000) + P$$

$$\$1,500,000 = P$$

Contribution margin to be earned in 2001:	
Total fixed cost ($500,000 – $80,000)	$ 420,000
Net income	1,500,000
Contribution margin needed for year	$1,920,000
Contribution margin already earned:	
(Selling price – variable cost) × units	
($25 – $15) $10 × 40,000 units	400,000
Contribution margin remaining	$1,520,000

$$\text{Contribution margin per unit needed} = \frac{\text{Contribution margin remaining}}{\text{Units remaining}}$$

$$= \frac{\$1,520,000}{110,000} = \$13.82$$

3.4 ARE YOU UTILIZING CAPACITY?

Contribution margin analysis can be used to determine the best way of utilizing capacity.

Example 3.9 — You can produce a raw metal that can either be sold at this stage or processed further and sold as an alloy.

	Raw Material	Alloy
Selling price	$200	$315
Variable cost	90	120

Total fixed cost is $400,000, and 100,000 hours of capacity are interchangeable between the products. There is unlimited demand for both products. Three hours are required to produce the raw metal, and five hours are needed to make the alloy. The contribution margin per hour is as follows:

	Raw Metal	Alloy
Selling price	$200	$315
Less: Variable cost	90	120
Contribution margin	$110	$195
Hours per ton	3	5
Contribution margin per hour	$ 36.67	$ 39

You should sell only the alloy because it results in the highest contribution margin per hour. Fixed costs are not considered because they are constant and are incurred irrespective of which product is manufactured.

3.5 CONCLUSION

Contribution margin analysis aids you in making sound departmental decisions. Is an order worth accepting even though it is below the normal selling price? What should the price of your product or service be? Is a proposed contract advantageous? What is your incremental profitability? What is the best way of using departmental capacity and resources?

In some cases, your bonus may be based on the contribution margin you earn for your department. Thus, an understanding of the computation of contribution margin is necessary.

4 Are You Breaking Even?

Before your business can realize "profit," you must first understand the concept of breaking even. To break even on your department's product lines and/or services, you must be able to calculate the sales volume needed to cover your costs and how to use this information to your advantage. You must also be familiar with how your costs react to changes in volume.

In an interview, a nonfinancial manager employed by a publishing company commented that:

> Our advertising percentage (paid advertisements appearing in a given issue) represents our break-even point. We always had to attain an advertising percentage of .67 or better. If the number fell below .67 and the paper was printed, it would cost the company money. The higher the percentage above .67, the more profit the company would make on a given issue. Thus, .67 is the paper's break-even point, because at this percentage our revenue from advertisers is equal to the costs incurred from printing and distributing the paper.

4.1 WHAT IS COST-VOLUME-PROFIT (CVP) ANALYSIS?

Cost-volume-profit (CVP) analysis relates to the way profit and costs change with a change in volume. CVP analysis examines the impact on earnings of changes in such factors as variable cost, fixed cost, selling price, volume, and product mix. CVP information helps you to predict the effect of any number of contemplated actions and to make better planning decisions. More specifically, CVP analysis tries to answer the following questions:

1. What sales volume is required to break even? How long will it take to reach that sales volume?
2. What sales volume is necessary to earn a desired profit?
3. What profit can be expected on a given sales volume?
4. How would changes in selling price, variable costs, fixed costs, and output affect profits?
5. How would a change in the mix of products sold affect the break-even and target volume and profit potential?

4.2 WHAT AND WHY OF BREAK-EVEN SALES

Break-even analysis, which is part of CVP analysis, is the process of calculating the sales needed to cover your costs so that there is zero profit or loss. The break-even point that is arrived at by such analysis is important to the profit planning process. Such knowledge allows managers to maintain and improve operating results. It is also important when introducing a new product or service, modernizing facilities, starting a new business, or appraising production and administrative activities.

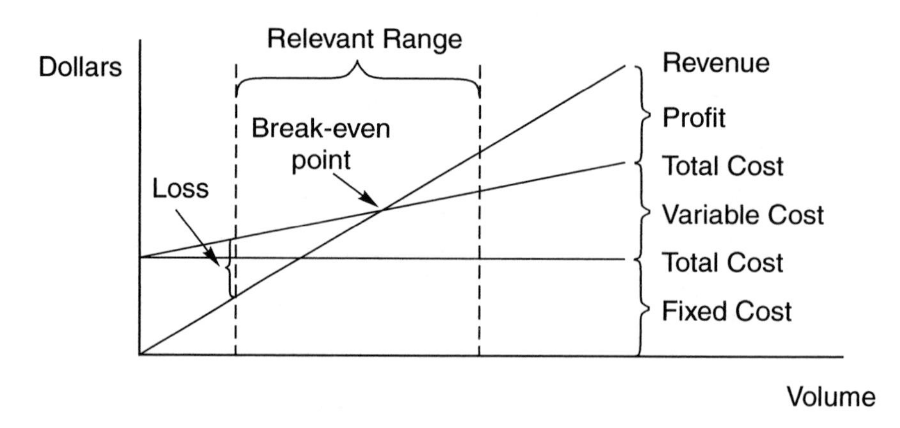

FIGURE 4.1 Break-even chart.

Break-even analysis can also be used as a screening device, such as the first attempt to determine the economic feasibility of an investment proposal.

Also, pricing may be aided by knowing the break-even point for a product. What other situations can you think of where break-even analysis is useful?

The assumptions of break-even analysis follow:

- Selling price is constant.
- There is only one product or a constant sales mix.
- Manufacturing efficiency is constant.
- Inventories do not significantly change from period to period.
- Variable cost per unit is constant.

The guidelines for breaking even are:

- An increase in selling price lowers break-even sales.
- An increase in variable cost increases break-even sales.
- An increase in fixed cost increases break-even sales.

Your objective, of course, is not just to break even, but to earn a profit. In deciding which products to push, continue, or discontinue, the break-even point is not the only important factor. Economic conditions, supply and demand, and the long-term impact on customer relations must also be considered. You can extend break-even analysis to concentrate on a desired profit objective.

The break-even sales can be determined using the graphic and equation approaches. Using the graphic approach (see Figure 4.1), revenue, total cost, and fixed cost are plotted on a vertical axis and volume is plotted on a horizontal axis. The break-even point occurs at the intersection of the revenue line and the total cost line. Figure 4.1 also depicts profit potentials over a wide range of activity. It shows how profits increase accordingly with increases in volume.

The equation approach uses the following equation:

$$S = \mathrm{VC} + \mathrm{FC}$$

where S = sales, VC = variable cost, and FC = fixed cost. This approach allows you to solve for break-even sales or for other unknowns as well. An example is selling price.

If you want a desired before-tax profit, solve for P in the following equation:

$$S = \text{VC} + \text{FC} + P$$

Example 4.1 — A product has a fixed cost of \$270,000 and a variable cost of 70% of sales. The point of break-even sales can be calculated as follows:

$$S = \text{FC} + \text{VC}$$

$$1S = \$270{,}000 + 0.7S$$

$$0.3S = \$270{,}000$$

$$S = \$900{,}000$$

If the selling price per unit is \$100, break-even units are 9,000 (\$900,000/\$100). If desired profit is \$40,000, the sales needed to obtain that profit (P) can be calculated as follows:

$$S = \text{FC} + \text{VC} + P$$

$$1S = \$270,\ 000 + 0.7S + \$40{,}000$$

$$0.3S = \$310{,}000$$

$$S = \$1{,}033{,}333$$

Example 4.2 — If the selling price per unit is \$30, the variable cost per unit is \$20, and the fixed cost is \$400,000, the break-even units (U) can be calculated as follows:

$$S = \text{FC} + \text{VC}$$

$$\$30U = \$400{,}000 + \$20U$$

$$\$10U = \$400{,}000$$

$$U = 40{,}000$$

The break-even dollar amount is:

$$40{,}000 \text{ units} \times \$30 = \$1{,}200{,}000$$

Example 4.3 — You sell 800,000 units of an item. The variable cost is \$2.50 per unit. Fixed cost totals \$750,000. The selling price (SP) per unit should be \$3.44 to break even:

$$S = \text{FC} + \text{VC}$$

$$800{,}000SP = \$750{,}000 + (\$2.50 \times 800{,}000)$$

$$800{,}000SP = \$2{,}750{,}000$$

$$SP = \$3.44$$

Example 4.4 — Assume your selling price is \$40, your sales volume is 20,000 units, your variable cost is \$15 per unit, your fixed cost is \$120,000, your after-tax profit is \$60,000, and your tax rate is 40%. To determine how much you have available to spend on research (R), consider this equation:

$$S = \text{VC} + \text{FC} + P + R$$

$$(\$40 \times 20{,}000) = (\$15 \times 20{,}000) + \$120{,}000 + \$100{,}000^{*} + R$$

$$\$280{,}000 = R$$

Example 4.5 — Assume your selling price is \$40, your variable cost is \$24, your fixed cost is \$150,000, your after-tax profit is \$240,000, and your tax rate is 40%. To determine how many units you must sell to earn the after-tax profit, consider the following equation:

$$S = \text{FC} + \text{VC} + P$$

$$\$40\ U = \$150{,}000 + \$24\ U + \$400{,}000^{**}$$

$$\$16\ U = \$550{,}000$$

$$U = 34{,}375$$

Example 4.6 — Assume your selling price is \$50 per unit, your variable cost is \$30 per unit, your sales volume is 60,000 units, your fixed cost is \$150,000, and your tax rate 30%. To determine the after-tax profit, use the following equation:

$$S = \text{FC} + \text{VC} + P$$

$$(\$50 \times 60{,}000) = 150{,}000 + (\$30 \times 60{,}000) + P$$

$$1{,}050{,}000 = P$$

$$\$1{,}050{,}000 \times 0.70 = \$735{,}000$$

---

* After-tax profit: \$60,000 = 0.6 × Before-tax profit

$$\frac{\$60{,}000}{0.6} = \text{Before-tax profit}$$

$$\$100{,}000 = \text{Before-tax profit}$$

** After-tax profit: \$240,000 = 0.6 × Before-tax profit

$$\frac{\$240{,}000}{0.6} = \text{Before-tax profit}$$

$$\$400{,}000 = \text{Before-tax profit}$$

Example 4.7 — You are considering making a product presently purchased outside for \$0.12 per unit. The fixed cost is \$10,000, and the variable cost per unit is \$0.08. Use the following equation to determine the number of units you must sell so that the annual cost of your machine equals the outside purchase cost.

$$\$0.12\ U = \$10{,}000 + \$0.08\ U$$

$$\$0.40\ U = \$10{,}000$$

$$U = 250{,}000$$

4.3 WHAT IS MARGIN OF SAFETY?

The *margin of safety* is a risk indicator that stipulates the amount by which sales may decline before losses are experienced.

$$\text{Margin of safety} = \frac{\text{Budget sales} - \text{Break-even sales}}{\text{Budget sales}}$$

The lower the ratio, the greater the risk of reaching the break-even point.

Example 4.8 — If budget sales are \$40,000 and break-even sales are \$34,000, what is your margin of safety?

$$\text{Margin of safety} = \frac{\$40{,}000 - \$34{,}000}{\$40{,}000} = 15\%$$

4.4 CASH BREAK-EVEN POINT

If you have a minimum of available cash, or if the opportunity cost of holding excess cash is high, you may want to know the volume of sales that will cover all cash expenses during a period. This is known as the *cash break-even point*.

Not all fixed costs involve cash payments. For example, depreciation expense is a noncash charge. To find the cash break-even point, the noncash charges must be subtracted from total fixed costs. Therefore, the cash break-even point is lower than the usual break-even point. The cash break-even point equation is as follows:

$$S = \text{VC} + \text{FC (after deducting depreciation)}$$

Example 4.9 — If the selling price is \$25 per unit, the variable cost is \$15 per unit, and total fixed cost is \$50,000, which includes depreciation of \$2,000, the cash break-even point is:

$$\$25\ U = \$15\ U + \$48{,}000$$

$$\$10\ U = \$48{,}000$$

$$U = 4{,}800$$

You must sell 4,800 units at \$25 each to meet your break-even point.

4.5 WHAT IS OPERATING LEVERAGE?

Operating Leverage is the degree to which fixed costs exist in your cost structure. It is the extent to which you commit yourself to high levels of fixed costs other than interest payments in order to leverage profits during good times. However, high operating leverage means risk because fixed costs cannot be decreased when revenue drops in the short run.

A high ratio of fixed cost to total cost over time may cause variability in profit. But a high ratio of variable cost to total cost indicates stability. It is easier to adjust variable cost than fixed cost when demand for your products decline.

Example 4.10 — Assume that fixed costs were $40,000 in 2000 and $55,000 in 2001, and that variable costs were $25,000 in 2000 and $27,000 in 2001.

The operating leverage in 2001 compared to 2000 was higher, as indicated by the increase in the ratio of fixed costs to total costs. Hence, there is greater earnings instability.

$$\frac{\$55{,}000}{\$82{,}000} = 67.1\% \text{ for } 2001$$

$$\frac{\$40{,}000}{\$65{,}000} = 61.5\% \text{ for } 2000$$

Example 4.11 — Assume that your selling price is $30 per unit, your variable cost is $18 per unit, your fixed cost is $40,000, and your sales volume is 8,000 units. You can determine the extent of operating leverage as follows:

$$\frac{(\text{Selling price} - \text{Variable cost})(\text{Units})}{(\text{Selling price} - \text{Variable cost})(\text{Units}) - \text{Fixed cost}}$$

$$\frac{(\$30 - \$18)(8{,}000)}{(\$30 - \$18)(8{,}000) - \$40{,}000} = \frac{\$96{,}000}{\$56{,}000} = 1.71$$

This means that for every 1% increase in sales above the 8,000-unit volume, income will increase by 1.71%. If sales increase by 10%, net income will rise by 17.10%.

Example 4.12 — You are evaluating operating leverage. Your selling price is $2 per unit, your fixed cost is $50,000, and your variable cost is $1.10 per unit. The first example assumes a sales volume of 100,000 units; the second assumes a sales volume of 130,000 units.

Sales Volume (in dollars)	–	Fixed Cost	–	Variable Cost	=	Net Income
(100,000 × $2)	–	$50,000	–	$110,000	=	$40,000
(130,000 × $2)	–	$50,000	–	$143,000	=	$67,000

The ratio of the percentage change in net income to the percentage change in sales volume is as follows:

$$\frac{\frac{\text{Change in net income}}{\text{Net Income}}}{\frac{\text{Change in quantity}}{\text{Quantity}}} = \frac{\frac{\$67,000 - \$40,000}{\$40,000}}{\frac{\$130,000 - \$100,000}{\$100,000}} = \frac{\frac{\$27,000}{\$40,000}}{\frac{\$30,000}{\$100,000}} = \frac{67.5\%}{30.0\%}$$

$$= 2.25$$

If fixed cost remains the same, the 2.25 figure tells you that for every 1% increase in sales above the 100,000-unit volume there will be a 2.25% increase in net income. Thus, a 10% jump in sales will boost net income 22.5%. The same proportionate operating leverage develops regardless of the size of the sales increase above the 100,000-unit level.

The opportunity to magnify the increase in earnings that arises from any increase in sales suggests that you should use a high degree of operating leverage. Presumably, fixed operating costs should constitute a larger proportion of the total cost at a particular sales level so as to enhance the gains realized from any subsequent rise in sales. While a high degree of operating leverage is sometimes a desirable objective, there is the risk of financial damage caused by a drop in sales. It should also be noted that the more unpredictable sales volume is, the more desirable it is to have a high degree of operating leverage. *Remember*: Fixed costs magnify the gain or loss from any fluctuation in sales.

If you have a high degree of operating leverage, you should not simultaneously use a high degree of financial leverage (debt) because the combination makes the risk associated with your operations too severe for a volatile economic environment. Alternatively, if you have a low degree of operating leverage you can often take on a higher level of financial leverage. The lower risk of operating leverage balances the higher risk of financial leverage (debt position). The tradeoffs determine how much financial and operating leverage to use.

One example of an operating leverage question you may face is whether to buy buildings and equipment or rent them. If you buy them, you incur fixed costs even if volume declines. If there is a rental with a *short-term* lease, the annual cost is likely to be more, but it is easier to terminate the fixed cost in a business downturn. Another operating leverage decision involves whether to purchase plant and facilities and manufacture all components of the product or to subcontract the manufacturing and just do assembly. With subcontracting, contracts can be terminated when demand declines. If plant and equipment are bought, the fixed cost remains even if demand declines.

Operating leverage is an issue that directly impacts line managers. The level of operating leverage selected should not be made without input from the production managers. In general, newer technology has a higher fixed cost and lower variable cost than older technology. Managers must determine whether the risks associated with higher fixed costs are worth the potential returns.

4.6 SALES MIX ANALYSIS

Break-even analysis requires some additional considerations when your department produces and sells more than one product. Different selling prices and different variable costs result in different unit contribution margins. As a result, break-even points vary with the relative proportions of the products sold, called the *sales* mix. In break-even analysis, it is necessary to predetermine the sales mix and then compute a weighted average contribution margin. It is also necessary to assume that the sales mix does not change for a specified period.

Example 4.13 — Your department has fixed costs of $76,000 and two products with the following contribution margin data:

	Product A	Product B
Selling price	$15	$10
Less: Variable cost	12	5
Unit contribution margin	$ 3	$ 5
Sales mix	60%	40%

The weighted average unit contribution margin is:

$3(.6) + $5(.4) = $3.80

Your department's break-even point in units is:

$76,000/$3.80 = 20,000 units

which is divided as follows:

Product A: 20,000 units × 60% = 12,000 units

Product B: 20,000 units × 40% = 8,000 units

Example 4.14 — Your department has total fixed costs of $18,600 and produces and sells three products:

	Product A	Product B	Product C	Total
Sales	$30,000	$60,000	$10,000	$100,000
Less: Variable cost	24,000	40,000	5,000	69,000
Contribution margin	$6,000	$20,000	$5,000	$31,000
Contribution margin ratio	20%	33.3%	50%	31%
Sales Mix	30%	60%	10%	100%

Since the contribution margin ratio for your department is 31%, the break-even point in dollars is:

$18,600/.31 = $60,000

which will be split in the mix ratio of 3:6:1 to give us the following break-even points for the individual products A, B, and C:

Product A:	\$60,000 × 30%	=	\$18,000
Product B:	\$60,000 × 60%	=	\$36,000
Product C:	\$60,000 × 10%	=	\$ 6,000
			\$60,000

One important assumption in a multiproduct department is that the sales mix will not change during the planning period. If the sales mix does change, however, the break-even point will also change.

4.7 CONCLUSION

If your sales volume consistently exceeds the break-even point, you may improve departmental profitability by reducing your variable cost of manual labor through a fixed cost investment in plant. However, you must consider the risk that sales may decline below the break-even point, which will decrease your profitability. Generally, you should aim to convert fixed cost to variable cost, rather than converting variable cost to fixed. But this must be done carefully, because it will reduce your contribution margin.

5 How to Make Short-Term, Nonroutine Decisions

When analyzing the manufacturing and/or selling functions of your department, you are faced with the problem of choosing between alternative courses of action. What should you produce? How should you manufacture it? Where should you sell the product? What price should you charge? In the short run, you may be confronted with the following nonroutine, nonrecurring situations:

- Whether to accept or reject a special order
- How to price standard products
- Whether to sell or process further
- Whether to make or buy
- Whether to add or drop a certain product line
- How to utilize scarce resources

5.1 WHAT COSTS ARE RELEVANT TO YOU?

In each of these situations, the ultimate management decision rests on cost data analysis. Cost data, which are classified by function, behavior patterns, and other criteria, are the basis for profit calculations. However, not all costs are of equal importance in decision making, and you must identify the costs that are relevant to a decision. The *relevant costs* are the future costs that differ between the decision alternatives. Therefore, the *sunk costs* are not relevant to these decisions because they are past costs. The *incremental or differential costs* are relevant because they are the ones that differ between the alternatives. For example, in a decision on whether to sell an existing business for a new one, the cost to be paid for the new business is relevant. However, the initial cost of the old business is not relevant because it is a sunk cost.

Under the concept of relevant costs, which may be appropriately titled *the incremental, differential, or relevant cost approach*, decision making involves the following steps:

1. Gather all costs associated with each alternative.
2. Drop the sunk costs.
3. Drop those costs that do not differ between alternatives.
4. Select the best alternative based on the remaining cost data.

Example 5.1 — You are planning to expand your department's productive capacity. The plan consists of purchasing a new machine for $50,000 and disposing of the old without receiving anything. The new machine has a 5-year life. The old machine has a 5-year remaining life and a book value of $17,500. The new machine will generate the same amount of revenue as the old one, but will substantially cut down on variable operating costs. Annual sales and operating costs of the present machine and the proposed replacement are based on normal sales volume of 20,000 units and are estimated as follows:

	Present Machine	New Machine
Sales	$60,000	$60,000
Less: Variable costs	35,000	20,000
Less: Fixed costs:		
Depreciation (straight-line)	2,500	10,000
Insurance, taxes, etc.	4,000	4,000
Net income	$18,500	$26,000

At first glance, it appears the new machine provides an increase in net income of $7,500 per year. However, sales and fixed costs such as insurance and taxes are irrelevant because they do not differ between the two alternatives. Eliminating all the irrelevant costs leaves only the incremental costs:

Savings in variable costs	$15,000
Less: Increase in fixed costs	10,000*
Net annual cash saving arising from the new machine	$5,000

* Exclusive of $2,500 sunk cost

You must also consider that the $50,000 cost of the new machine is relevant; however, the book value of the present machine is a sunk cost and is irrelevant in this decision.

5.2 ACCEPTING OR REJECTING A SPECIAL ORDER

You may receive a short-term, special order for your products at a lower price than usual. Normally, you may refuse such an order since it will not yield a satisfactory profit. However, if sales are slumping, such an order should be accepted if the incremental revenue obtained from it exceeds the incremental costs involved. The company should accept this price since it is better to receive some revenue than to receive nothing at all. A price that is lower than the regular price is called a *contribution price*. This *contribution approach to pricing* is most appropriate when: (1) there is a distressing operating situation where demand has fallen off; (2) there is idle capacity; or (3) there is sharp competition or a competitive bidding situation.

Example 5.2 — Your department has a 100,000-unit capacity. You are producing and selling only 90,000 units of a product each year at a regular price of $2. If the variable cost per unit is $1 and the annual fixed cost is $45,000, the income statement follows:

	Total	Per Unit
Sales (90,000 units)	$180,000	$2.00
Less: Variable costs (90,000 units)	90,000	1.00
Contribution margin	$90,000	$1.00
Less: Fixed cost	45,000	0.50
Net income	$45,000	$0.50

The company received an order calling for 10,000 units at $1.20 per unit, for a total of $12,000. The buyer will pay the shipping expenses. Although the acceptance of this order will not affect regular sales, you are reluctant to accept it because the $1.20 price is below the $1.50 factory unit cost ($1.50 = $1.00 + $0.50). You must consider, however, that you can add to total profits by accepting this special order even though the price offered is below the unit factory cost. At a price of $1.20, the order will contribute $0.20 per unit (contribution margin per unit = $1.20 – $1.00 = $0.20) toward fixed cost, and profit will increase by $2,000 (10,000 units × $0.20). Using the contribution approach to pricing, the variable cost of $1 will be a better guide than the full unit cost of $1.50. Note that the fixed costs will not increase.

	Per Unit	Without Special Order (90,000 Units)	With Special Order (100,000 Units)	Difference
Sales	$2.00	$180,000	$192,000	$12,000
Less: Variable costs	1.00	90,000	100,000	10,000
Contribution margin	$1.00	$ 90,000	$ 92,000	$ 2,000
Less: Fixed cost	0.50	45,000	45,000	
Net income	$0.50	$45,000	$47,000	$2,000

5.3 PRICING STANDARD PRODUCTS

Unlike pricing special orders, pricing standard products requires long-term considerations. The key concept is to recognize that the established unit selling price must be sufficient in the long run to cover all manufacturing, selling, and administrative costs, both fixed and variable, as well as to provide for an adequate return and for future expansion. There are two primary approaches to pricing standard products that are sold on the regular market, both of which use some kind of cost-plus pricing formula.

1. The *full-cost approach* defines the cost base as the full unit manufacturing cost. Selling and administrative costs are provided for through the markup that is added to the cost base.
2. The *contribution approach* defines the cost base as the unit variable cost. Fixed costs are provided for through the markup that is added to this base.

Example 5.3 — You have prepared the following cost data on your company's regular product:

	Per Unit	Total
Direct material	$6	
Direct labor	4	
Variable overhead	4	
Fixed overhead (based on 20,000 units)	6	$120,000
Variable selling and administrative expenses	1	
Fixed selling and administrative expenses (based on 20,000 units)	2	$40,000

Assume that in order to obtain its desired selling price, the firm has a general policy of adding a markup equal to 50% of the full unit cost or 100% of the unit variable cost.

Under the full-cost approach, the desired unit selling price is:

Direct material	$6
Direct labor	4
Factory overhead	10
Full unit cost	$20
Markup to cover selling and administrative expenses, and desired profit - 50% of full unit cost	10
Desired selling price	$30

Under the contribution approach, the desired selling price is determined as follows:

Direct material	$6
Direct labor	4
Variable costs (overhead, selling, and administrative expenses)	5
Unit variable cost	$15
Markup to cover fixed costs, and desired profit - 100% of unit variable cost	15
Desired selling price	$30

Example 5.4 — Your company has determined that a $500,000 investment is necessary to manufacture and market 20,000 units of its product every year. It will cost $20 to manufacture each unit at a 20,000-unit level of activity, and total selling and administrative expenses are estimated to be $100,000. If your company desires a 20% return on investment, what will be the markup using the full-cost approach?

Desired rate of return (20% × $500,000)	$100,000
Selling and administrative expenses	100,000
Total	$200,000
Full unit cost (20,000 units × $20)	$400,000
Required markup (Total/Full unit cost)	50%

5.4 ANALYZING THE MAKE-OR-BUY DECISION

Deciding whether to produce a component part internally or to buy it from a supplier is called a *make-or-buy decision.* This decision involves both qualitative factors (e.g., product quality and long-term business relationships with subcontractors) and quantitative factors (e.g., cost). The quantitative effects of the make-or-buy decision are best seen through incremental analysis.

Example 5.5 — You have prepared the following cost estimates for manufacture of a subassembly component based on an annual product of 8,000 units:

	Per Unit	Total
Direct material	$ 5	$40,000
Direct labor	4	32,000
Variable factory overhead applied	4	32,000
Fixed factory overhead applied (150% of direct labor cost)	6	$48,000
Total Cost	$19	$152,000

A supplier offers the subassembly at a price of $16 each. Two thirds of fixed factory overhead, which represent executive salaries, rent, depreciation, and taxes, continue regardless of the decision. To determine whether to make or buy the product, you must evaluate the relevant costs that can change between the alternatives. Assuming productive capacity will be idle if not used to produce the subassembly, the analysis is as follows:

	Per Unit		Total of 8,000 Units	
	Make	Buy	Make	Buy
Purchase price		16		128,000
Direct material	$5		$ 40,000	
Direct labor	4		32,000	
Variable overhead	4		32,000	
Fixed overhead that can be avoided by not making	2		16,000	
Total relevant costs	$15	$16	$120,000	$128,000
Difference in favor of making		$ 1		$ 8,000

The make-or-buy decision must be evaluated in the broader perspective of considering how best to utilize available facilities. The alternatives include:

1. Leaving facilities idle;
2. Renting out idle facilities; or
3. Using idle facilities for other products.

5.5 DETERMINING WHETHER TO SELL OR PROCESS FURTHER

When two or more products are produced simultaneously from the same input by a joint process, these products are called *joint products*. The term *joint costs* is used to describe all the manufacturing costs incurred prior to the point at which the joint products are identified as individual products, that is, the *split-off point*. At the split-off point some of the joint products are in final form and can be sold to the consumer, whereas others require additional processing. In many cases, however, you might have an option: you can sell the goods at the split-off point or process them further in the hope of obtaining additional revenue. Joint costs are considered irrelevant to this sell-or-process-further decision, since the joint costs have already been incurred at the time of the decision and, therefore, represent sunk costs. The decision will rely exclusively on additional revenue compared to the additional costs incurred due to further processing.

Example 5.6 — Your company produces products A, B, and C from a joint process. Joint production costs for the year are $120,000. Product A may be sold at the split-off point or processed further. The additional processing requires no special facilities, and all additional processing costs are variable. The sales value at the split-off point of 3,000 units is $60,000. The sales value for 3,000 units after further processing is $90,000 and the additional processing cost is $25,000.

Incremental sales revenue	$30,000
Incremental costs of additional processing	25,000
Incremental revenue of additional processing	5,000

It is profitable for product A to be processed further. Keep in mind that the joint production cost of $120,000 is not included in the analysis, since it is a sunk cost and, therefore, is irrelevant to the decision.

5.6 ADDING OR DROPPING A PRODUCT LINE

Deciding whether to drop an old product line or add a new one requires an evaluation of both qualitative and quantitative factors. However, any final decision should be based primarily on the impact on contribution margin or net income.

Example 5.7 — Your department has three major product lines: A, B, and C. You are considering dropping product line B because it is being sold at a loss. The income statement for these product lines follows:

	Product A	Product B	Product C	Total
Sales	$10,000	$15,000	$25,000	$50,000
Less: Variable costs	6,000	8,000	12,000	26,000
Contribution margin	4,000	7,000	13,000	24,000

	Product A	Product B	Product C	Total
Less: Fixed costs				
Direct	2,000	6,500	4,000	12,500
Allocated	1,000	1,500	2,500	5,000
Total	3,000	8,000	6,500	17,500
Net income	$ 1,000	$(1,000)	$ 6,500	$ 6,500

Direct fixed costs are identified directly with each of the product lines, whereas allocated fixed costs are common fixed costs allocated to the product lines using some base (e.g., space occupied). Common fixed costs typically continue regardless of the decision and thus cannot be saved by dropping the product line to which they are distributed.

The following calculations show the effects on your department with and without product line B.

	Keep Product B	Drop Product B	Difference
Sales	$50,000	$35,000	$(15,000)
Less: Variable cost	26,000	18,000	(8,000)
Contribution margin	24,000	17,000	(7,000)
Less: Fixed costs			
Direct	12,500	6,000	(6,500)
Allocated	5,000	5,000	
Total	17,500	11,000	(6,500)
Net income	$ 6,500	$ 6,000	$ 500

Alternatively, if product line B were dropped, the incremental approach would show the following:

Sales revenue lost		$15,000
Gains:		
Variable cost avoided	$8,000	
Directed fixed costs avoided	6,500	
	14,500	14,500
Increase (decrease) in net income		$ (500)

Both methods demonstrate that by dropping product line B your department will lose an additional $500. Therefore, product line B should be kept. One of the great dangers in allocating common fixed costs is that such allocations can make a product line look less profitable than it really is. Because of such an allocation, product line B showed a loss of $1,000 but it actually contributes $500 ($7,000 - $6,500) to the recovery of common fixed costs.

5.7 UTILIZING SCARCE RESOURCES

In general, the emphasis on products with higher contribution margins maximizes your department's net income. This is not true, however, when there are constraining factors or scarce resources. A *constraining factor* restricts or limits the production or sale of a given product. It may be machine hours, labor hours, or cubic feet of warehouse space. In the presence of these constraining factors, maximizing profit depends on getting the highest contribution margin per unit of the *factor* (rather than the highest contribution margin per unit of *product output*).

Example 5.8 — Your department produces products A and B with the following contribution margins per unit and an annual fixed cost of $42,000.

	Product A	Product B
Sales	$8	$24
Less: Variable costs	6	20
Contribution margin	$2	$ 4

As is indicated by the contribution margin, product B is more profitable than product A since it contributes more to your department's profits ($4 vs. $2). But assume that your department has a limited capacity of 10,000 labor hours. Further, assume that product A requires 2 labor hours to produce and product B requires 5 labor hours. One way to express this limited capacity is to determine the contribution margin per labor hour.

	Product A	Product B
Contribution margin per unit	$2.00	$4.00
Labor hours required per unit	2	5
Contribution margin per labor hour	$1.00	$0.80

Since product A returns the higher contribution margin per labor hour, it should be produced and product B should be dropped.

Another way to look at the problem is to calculate the total contribution margin for each product.

	Product A	Product B
Maximum possible production	5,000 units*	2,000 units**
Contribution margin per unit	× $2	× $4
Total contribution margin	$10,000	$8,000

* (10,000 hours/2 hours)
** (10,000 hours/5 hours)

Again, product A should be produced since it contributes more than product B ($10,000 vs. $8,000).

5.8 DO NOT FORGET THE QUALITATIVE FACTORS

In addition to the quantitative factors, incremental qualitative factors must also be considered in the decision making. Qualitative factors, which are difficult to measure in terms of money, include:

1. Effect on employee morale, schedules, and other internal factors;
2. Relationships with and commitments to suppliers;
3. Effect on present and future customers; and
4. Long-term future effect on profitability.

In some decision-making situations, qualitative factors are more important than immediate financial benefit.

5.9 CONCLUSION

Not all costs are of equal importance in decision making. The relevant costs are the expected future costs that differ between the decision alternatives. Therefore, the sunk costs are irrelevant since they are past costs. The costs that continue regardless of the decision are also irrelevant. The relevant cost approach assists you in making short-term, nonroutine decisions such as whether to accept a below-normal selling price, which products to emphasize, whether to make or buy, whether to sell or process further, and how to optimize utilization of capacity. In addition to these quantitative aspects, remember that qualitative factors must also be considered in many decision-making situations.

6 Financial Forecasting and Budgeting

To accomplish your department's goals you must be familiar with financial forecasting and budget planning. Without these instruments you have no plan on which to base future financial decisions, and without a plan you cannot make things happen. As a nonfinancial manager, you may have to forecast sales, production, and costs under varying assumptions. Learning to prepare and use financial forecasts and budgets will help you to accurately plan and control your department's future.

6.1 WHAT IS A FORECAST?

Despite the uncertainties in life, it is important to attempt to forecast what will happen in the future. A forecast is a starting point for planning. It may simply be a projection that what happened last year will happen this year, or it may be quite complex, using statistical techniques and seasonal patterns. Although you can base your projection on historical patterns, you must also take into account any important changes taking place in the current environment, as these changes will make the future different from the past. Examples of relevant changes are better technology, changes in laws and regulations, and increased competition.

Forecasts are the basis for capacity planning, inventory planning, planning for sales and market share, and budgeting. Sales forecasts are particularly important in many financial activities, including budgets, profit planning, capital expenditure analysis (whether to buy long-term assets), and acquisition and merger analysis (when two companies combine into one).

You must realize that forecasts are guesses regarding the future. The guesses principally arise from hard data rather than from subjective judgment; therefore, certain variables may not be accounted for in this process. Hence, you should question forecasted figures that seem unreasonable and illogical based on your experience. After all, you are most familiar with the financial factors that relate to your job.

6.2 HOW CAN YOU USE FORECASTS?

Forecasts are needed for marketing, production, manpower, and financial planning and are used for planning and implementing long-term objectives.

If you are a marketing manager, for instance, you can use sales forecasts to determine optimal sales force allocations, establish sales goals, and plan promotions and advertising. If you are a production planner, you need forecasts to schedule production activities, order materials, and plan shipments. If you are a purchaser, you need forecasts to aid you in determining what quantity of merchandise is needed and when to satisfy projected sales.

Figure 6.1 illustrates the relationship between forecasting and managerial functions. For example, as soon as your company is sure that it has sufficient capacity, the production plan is developed. If the company does not have enough capacity, planning and budgeting decisions will be required for capital spending to expand capacity.

In planning for capital investments, predictions of economic activity are required so that the returns, or cash inflows, accruing from the investment may be estimated. Borrowing needs for future operations should also be estimated through forecasts. Financial managers must make forecasts of money and credit conditions and interest rates so that the cash needs of the firm may be met at the lowest possible cost (e.g., should the company issue stock or debt to maintain its desired financial structure?). Long-term forecasts are needed to plan changes in the company's capital structure.

6.3 HOW DO YOU PREPARE A FINANCIAL FORECAST?

To prepare a financial forecast you must begin by forecasting sales. Production and related cost forecasting will follow. Forecasting is easiest when stability exists in relationships that can be captured in mathematical terms (e.g., $s = a + bA + cp$ where s = sales, A = advertising expenses and p = price).

The sequence for forecasting financing needs is:

1. Project your sales.
2. Project variables such as cost of sales and operating expenses.
3. Estimate the level of investment in assets required to support the projected sales.
4. Calculate financing needs.
5. Consider external factors affecting the forecast. Political concerns, economic environment, tax law, technological developments, and demographic factors may be applicable in certain cases.

Most other forecasts (budgets) follow the sales forecast. Sales estimates may be based on several factors including your instincts, past experiences, feedback from the sales force, and consumer surveys. If any changes are expected in your product mix (lines added or deleted), the projected change in volume is a consideration in projecting sales. If competition or cost reductions necessitate price changes, these must also be considered in making your sales forecast. A sales forecast of major product lines for a 6-month period is shown in Figure 6.2.

6.4 PERCENT-OF-SALES METHOD OF FINANCIAL FORECASTING

When constructing a financial forecast, the sales forecast is used to estimate various expenses, assets, and liabilities. Often, sales forecasting is done by product line. The

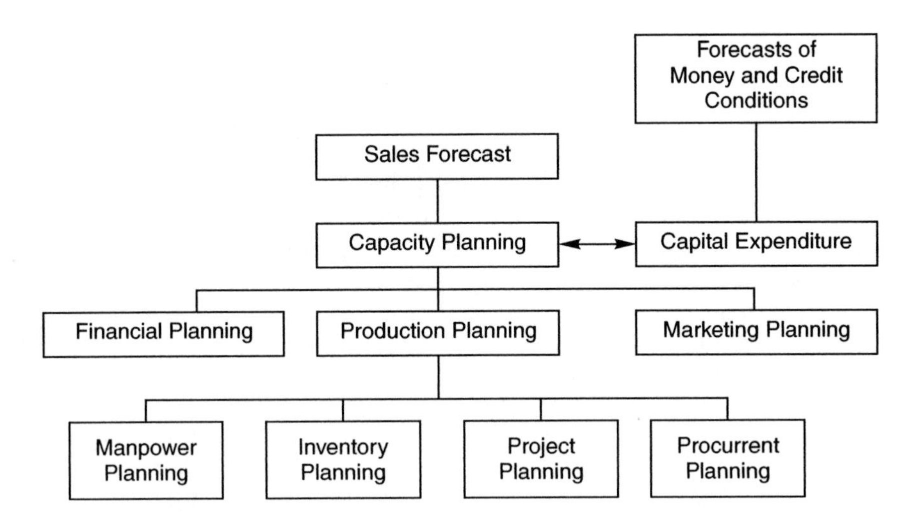

FIGURE 6.1 Sales forecasts and managerial functions.

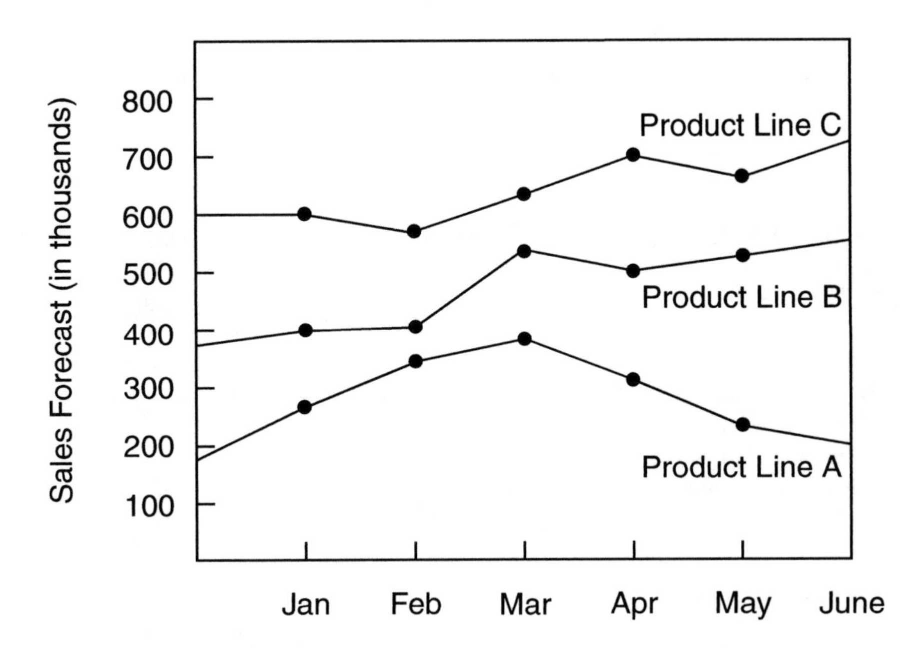

FIGURE 6.2 Sales forecast by product line.

most widely practiced method for formulating projections is the *percent-of-sales method*, in which the various expenses, assets, and liabilities for a future period are estimated as a percentage of sales for the present period. These percentages, together with the projected sales for the upcoming year, are then used to construct *pro forma* (planned or projected) balance sheets.

The computations for a *pro forma* balance sheet follow:

1. Express balance sheet items that *vary directly with sales* as a percentage of sales. Any item that does not vary directly with sales (e.g., long-term debt) is considered not applicable (n.a.).
2. Multiply the percentages determined in Step 1 by the projected sales figure to obtain the amounts for the future period.
3. Where no percentage applies (e.g., for long-term debt, common stock, and paid-in-capital), simply insert the figures from the present balance sheet in the column for the future period.
4. Compute the projected retained earnings as follows:

   Projected retained earnings = present retained earnings + projected net income – cash dividends

   (You will need to know the percentage of sales that represents net income and the dividend payout ratio.)
5. Sum the asset accounts to obtain a total projected assets figure. Then add the projected liabilities and equity accounts to determine the total financing provided. Since liabilities plus equity must balance the assets when totaled, any difference is a *shortfall*, which is the amount of external financing required.

Let us numerically work one out.

> **Example 6.1** — For the *pro forma* balance sheet shown in Table 6.1, net income is assumed to be 5% of sales and the dividend payout ratio is 40%.

One important limitation of the percent-of-sales method is that your company is assumed to be operating at full capacity. Based on this assumption, the firm does not have adequate productive capacity to absorb projected increases in sales and may require an additional investment in assets.

The prime advantage of the percent-of-sales method of financial forecasting is that it is simple and inexpensive. To obtain a more precise projection of future financing needs, however, the preparation of a cash budget is required.

6.5 WHAT IS A BUDGET?

In a personal interview with the author, a nonfinancial manager discussed the budgeting process as follows:

> Budgeting at the firm follows the traditional top-down planning strategy. Sales budgets are set up for the firm as a whole, then broken out into the four sales regions — eastern, central, western, and control states. At the Eastern Region we then receive our share of the corporate budget, which we must then subdivide among the territories in the region. The two key budgets that must be given considerable thought in their divisions are the

TABLE 6.1
***Pro Forma* Balance Sheet in Millions of Dollars**

	Present (2000)	% of Sales (2000 Sales = $20)	Projected (2001 Sales = $24)	
Assets				
Current	2	10	2.4	
Fixed	4	20	4.8	
Total assets	6		7.2	
Liabilities and stockholders' equity				
Current liabilities	2	10	2.4	
Long-term debt	2.5	n.a.	2.5	
Total liabilities	4.5		4.9	
Common stock	0.1	n.a.	0.1	
Paid-in-capital	0.2	n.a.	0.2	
Retained earnings	1.2		1.92	
Total equity	1.5		2.22	
Total liabilities and stockholders' equity	6		7.12	Total financing provided
			0.08	External financing needed
			7.2	Total

2001 retained earnings = 2000 retained earnings + projected net income – cash dividends paid

= $1.2 + 5%($24) – 40%[5%($24)]

= $1.2 + $1.2 – $0.48 = $2.4 – $0.48 = $1.92

External financing needed = projected total assets – (projected total liabilities + projected equity)

= $7.2 – ($4.9 + $2.22) = $7.2 – $7.12 = $0.08

sales quotas themselves and the promotional local funding dollars. Travel and expense dollars for the field representatives must be allocated. On a less formal, periodic basis, we receive allocations of promotional items that are charged to a national budget or charged in whole to our region and also must be allocated to the territories in the region.

In a personal interview with the author, another nonfinancial manager commented:

We use our projected sales forecasts for determining our increase in [insurance] policies in force. I am directly involved in our direct labor budget as we plan out our total people needs. I also must budget what each of my people will cost the company in terms of salary expense. I must budget for a certain percentage of promotions and also budget for the dollar amount and frequency of merit increases. This along with other aspects combine to formulate our yearly plan. After that is established we have everything in terms of variances to plan. Our management staff is then evaluated on our results in terms of variances. I am responsible for overtime dollars, additions to staff, and our in-house efficiencies in terms of production plans.

Budgets are very important to nonfinancial managers. They are useful for planning your goals, evaluating your progress, motivating your people, and communicating your results. Often your attainment of budget goals is measured in terms of variances. Budgeting is also very useful on a personal level, as you always want to know where you are in relation to your goals.

A *budget* is an annual projection of what you expect to achieve financially and operationally. It is a plan that commits resources to activities or projects and makes formal (written) statements of your expectations of sales, expenses, production, volume, and various financial transactions for the coming period A budget places guidelines and limitations on spending. The most common budget types that a nonfinancial manager might have to deal with are fixed budgets and flexible budgets. A fixed budget allocates a specific amount of resources that can be used to complete a specific task — this amount cannot be exceeded. A flexible budget allows the amount of allocated resources to vary along with the various levels of activity on a specific project.

Almost all nonfinancial managers are responsible for either planning and submitting a budget for a specific period of time or for conforming to one that has already been laid out.

In a personal interview, a nonfinancial manager employed by a computer company commented:

> Although the company's attitude toward the budgeting process is a bottom-up approach, the calculation of the budget is almost entirely a top-down method. My immediate superior hands me the annual budget and discusses it with me for any discrepancies.
>
> Our basic strategy to budgeting can be viewed as follows: first, strategic aspects are analyzed for both the short and long term; next, specific plans are developed; and last, the actual budgets are formulated.

The budget coordinates the activities of all the departments in your company. It communicates companywide goals to nonfinancial managers in different departments so they know what direction to take. You have no choice but to be familiar with your departmental budget and determine ways to keep within it. Many nonfinancial managers also believe that they must use all the money allocated to them in the budget, so as not to lose funding next year; however, this does not mean that funds should be spent foolishly just to meet the budget.

A budget should be realistic (e.g., feasible in marketing and production capabilities) and consistent with company policy. It should take into consideration what your department's or company's *mission* is. For instance, does your company want to hold the line against the competition or does it want to grow? Should your company take an aggressive or a conservative approach in achieving these goals? The budget should be constantly evaluated since external or internal conditions may warrant changes. A budget is only an estimate, and deviation from that estimate should be expected. The budget must be adaptable to changing circumstances in your department. What will happen to sales if conditions change? What problems will be caused if initial sales expectations prove wrong? It is best to develop cost figures for different sales levels (e.g., high, low, expected) to assure flexibility.

The budget is a blueprint for planning and control. It enables you to see your role and set goals accordingly. At the beginning of the period, the budget is a standard. If you have no initial plan, you have no idea of what you want to accomplish. At the end of the period, the budget is a control device to measure your department's performance against the plan so that future performance may be improved. If you do not know where you are going, it is hard to tell whether or not you have arrived. The budgetary time period depends on many factors, including how crucial the budget is to control costs. The period for a budget can be any appropriate time frame such as yearly, quarterly, monthly, weekly, and daily.

The major steps in preparing the budget are: (1) completing a sales forecast; (2) determining expected production volume; (3) estimating manufacturing costs and operating expenses; (4) determining cash flow and other financial effects; and (5) preparing projected financial statements. The budgetary process involves establishing goals, developing strategies, noting changes in the customer base, formulating plans of action, evaluating the market, looking at economic and political conditions, reviewing suppliers, analyzing competition, appraising industry trends, identifying the life cycle of the product, reviewing technological changes, appraising financial strength, and taking corrective action. However, without competent management, budgeting is a waste of time and money. A manager with a poor understanding of budgets may fail to correctly anticipate the future and commit to objectives for which his or her budget is inadequate, which may result in a budget overrun.

6.6 WHAT ASSUMPTIONS MUST BE MADE?

Prior to the budget being developed, certain questions must be asked and certain assumptions must be made. What will the inflation rate be? Where is competition headed? Will suppliers increase prices? Will customer tastes change? You also must explore the financial alternatives available to you. For instance, what will occur if you raise your selling price? What will be the effect if one variable (e.g., advertising) is changed?

6.7 WHAT IS THE STRUCTURE OF THE BUDGET?

The master (comprehensive) budget contains a projection of each of the key financial statements. Supporting schedules are also prepared for each of the major components in the financial statements. The master budget is classified broadly into two categories — the *operational budget* and the *financial budget*. The operational budget reflects the results of operating decisions and provides data needed to prepare a budgeted income statement. The financial budget shows the financial decisions of your company and includes the cash budget and financial statement projections other than the income statement. A simplified diagram of the various components of the comprehensive budget is shown in Figure 6.3.

The operational budget includes the following:

- The sales budget
- The direct materials budget

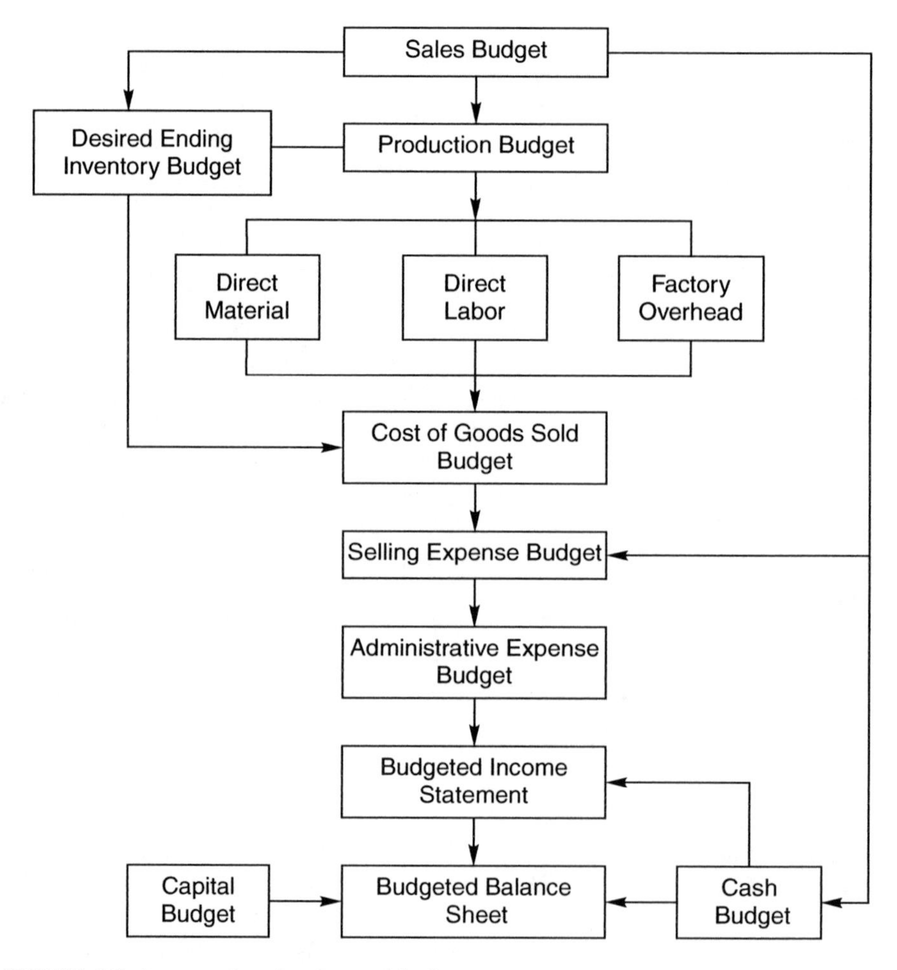

FIGURE 6.3 A comprehensive (master) budget.

- The direct labor budget
- The factory overhead budget
- The selling and administrative expense budget
- The *pro forma* income statement

The financial budget includes the following:

- The cash budget
- The *pro forma* balance sheet

To illustrate how all these budgets are put together, we will focus on a manufacturing company called the Johnson Company, which produces and markets a single product. We will assume that the company develops the master budget in contribution format for 2001 on a quarterly basis. We will highlight the variable cost-fixed cost breakdown throughout this example.

6.7.1 The Sales Budget

The sales budget is the starting point in preparing the master budget, since estimated sales volume influences nearly all other items appearing in the master budget. The sales budget, which ordinarily indicates the quantity of each product expected to be sold, allows all departments to plan their needs. Further, sales forecast figures determine staffing needs to reach targeted goals.

Accurate sales forecasting is important since an exaggerated sales forecast may result in your company's hiring too many staff members or acquiring excessive facilities, which translates into unneeded costs. On the other hand, a very pessimistic forecast may cause a shortage in staff and facilities so that consumer demand cannot be met; this translates into lost business. Nonfinancial managers must also carefully appraise the sales estimates of other departments. For instance, the marketing department may give optimistic sales estimates because it may try to stimulate the sales effort. As a result, it may not realistically appraise consumer demand.

After sales volume has been estimated, the sales budget is constructed by multiplying the expected sales in units by the expected unit sales price. Generally, the sales budget includes a computation of expected cash collections from credit sales, which will be used later for cash budgeting.

Example 6.2 — The Johnson Company's sales budget for the year ending December 31, 2001 is as follows:

	Quarter				
	1	2	3	4	Total
Expected sales in units	800	700	900	800	3,200
Unit sales price	× $80	× $80	× $80	× $80	× $80
Total sales	$64,000	$56,000	$72,000	$64,000	$256,000

The schedule of expected cash collections for the Johnson Company is as follows:

	Quarter				
	1	2	3	4	Total
Accounts receivable (12/31/2000)	$ 9,500[a]				$ 9,500
First quarter sales ($64,000)	44,800[b]	$17,920[c]			62,720
Second quarter sales ($56,000)		39,200[b]	$15,680[c]		54,880
Third quarter sales ($72,000)			50,400[b]	$20,160[c]	70,560
Fourth quarter sales ($64,000)				44,800[b]	44,800
Total cash collections	$54,300	$57,120	$66,080	$64,960	$242,460

[a] All of the $9,500 accounts receivable balance is assumed to be collected in the first quarter.

[b] Of the total quarter's sales, 70% are collected in the quarter of sale.

[c] Of the total quarter's sales, 28% are collected in the quarter following; the remaining 2% are uncollectible.

6.7.2 The Production Budget

After sales are budgeted, the production budget can be determined. The number of units expected to be manufactured to meet budgeted sales and inventory requirements (determined by subtracting the estimated inventory at the beginning of the period from the sum) is set forth in the production budget. The expected volume of production is of the units expected to be produced and the desired inventory at the end of the period.

Good managers constantly evaluate whether production is on schedule with proposed budgets and make adjustments accordingly. For example a computer or an assembly line is down for an extended period of time; the attentive manager knows that production and output will probably be low for that time period.

Example 6.3 — The Johnson Company's production budget for the year ending December 31, 2001 is as follows:

	Quarter				
	1	2	3	4	Total
Planned sales (see Example 6.2)	800	700	900	800	3,200
Desired ending inventory[a]	70	90	80	100[b]	100[c]
Total needs	870	790	980	900	3,300
Less: Beginning inventory[d]	80	70	90	80	80
Units to be produced	790	720	890	820	3,220

[a] This figure is calculated as 10% of the next quarter's sales.
[b] Estimated.
[c] The ending inventory for the year is the same as the one for the fourth quarter.
[d] The same as the previous quarter's ending inventory.

6.7.3 The Direct Material Budget

When the level of production has been computed, a direct material budget should be constructed to show how much material will be required for production and how much material must be purchased to meet this production requirement. The purchase will depend on both expected usage of material and inventory levels. The formula for computation of the purchase is as follows:

Purchase in units = Usage + Desired ending material inventory units
– Beginning inventory units

The direct material budget is usually accomplished by a computation of expected cash payments for materials.

Example 6.4 — The Johnson Company's direct material budget for the year ending December 31, 2001 is as follows:

	Quarter				
	1	2	3	4	Total
Units to be produced (see Example 6.3)	790	720	890	820	3,220
Material needs per unit (in lbs.)	×3	×3	×3	×3	×3
Material needs for production (in lbs.)	2,370	2,160	2,670	2,460	9,660
Desired ending inventory of materials (in lbs.)[a]	216	267	246	250[b]	250
Total material needs (in lbs.)	2,586	2,427	2,916	2,710	9,910
Less: Beginning inventory of materials (in lbs.)[c]	237	216	627	246	237
Materials to be purchased	2,349	2,211	2,649	2,464	9,673
Unit price	× $2	× $2	× $2	× $2	× $2
Purchase cost	$4,698	$4,422	$5,298	$4,928	$19,346

[a] 10% of the next quarter's units needed for production multiplied by 3 (material needs per unit).
[b] Estimated.
[c] The same as the prior quarter's ending inventory.

The schedule of expected cash disbursements for the Johnson Company is as follows:

	Quarter				
	1	2	3	4	Total
Accounts payable (12/31/2000)	$2,200				$2,200
First quarter purchases ($4,698)	2,349[a]	$2,349[b]			4,698
Second quarter purchases ($4,422)		2,211[a]	$2,211[b]		4,422
Third quarter purchases ($5,298)			2,649[a]	$2,649[b]	5,298
Fourth quarter purchases ($4,928)				2,464[a]	2,464
Total disbursements	$4,549	$4,560	$4,860	$5,113	$19,082

[a] Of the total quarter's purchases, 50% are paid for in the quarter of purchases.
[b] The remaining 50% of a quarter's purchases are paid for in the following quarter.

6.7.4 The Direct Labor Budget

The production requirements provide the starting point for the direct labor budget. To compute direct labor requirements, expected production volume for each period is multiplied by the number of direct labor hours to produce a single unit. The direct labor hours required to meet production needs are then multiplied by the direct labor cost per hour to obtain budgeted total direct labor costs.

Example 6.5 — The Johnson Company's direct labor budget for the year ending December 31, 2001 is as follows:

	Quarter				
	1	2	3	4	Total
Units to be produced (see Example 6.3)	790	720	890	820	3,200
Direct labor hours per unit	×5	×5	×5	×5	×5
Total hours	3,950	3,600	4,450	4,100	16,100
Direct labor cost per hour	× $5	× $5	× $5	× $5	× $5
Total direct labor cost	$19,750	$18,000	$22,250	$20,500	$80,500

6.7.5 The Factory Overhead Budget

The factory overhead budget provides a schedule of manufacturing costs other than direct materials and direct labor. These costs include factory rent, factory insurance, factory property taxes, and factory utilities. Using the contribution approach to budgeting requires the development of a predetermined overhead rate for the variable portion of the factory overhead. In developing the cash budget, remember that depreciation is not a cash outlay; therefore, it must be deducted from the total factory overhead in computing cash disbursement for factory overhead.

Example 6.6 — To illustrate the factory overhead budget, we will assume the following:

- Total factory overhead budgeted = \$6,000 fixed (per quarter) plus \$2 per hour of direct labor.
- Depreciation expenses are \$3,250 each quarter.
- All overhead costs involving cash outlays are paid for in the quarter incurred.

The Johnson Company's factory overhead budget for the year ending December 31, 2001 is as follows:

	Quarter				
	1	2	3	4	Total
Budgeted direct labor hours (see Example 6.5)	3,950	3,600	4,450	4,100	16,100
Variable overhead rate	× \$2	× \$2	× \$2	× \$2	× \$2
Variable overhead budgeted	7,900	7,200	8,900	8,200	32,200
Fixed overhead budgeted	6,000	6,000	6,000	6,000	24,000
Total budgeted overhead	13,900	13,200	14,900	14,200	56,200
Less: Depreciation	3,250	3,250	3,250	3,250	13,000
Cash disbursement for overhead	10,650	9,950	11,650	10,950	43,200

6.7.6 The Ending Inventory

The desired ending inventory budget provides information for the construction of budgeted financial statements. Specifically, it will help compute the cost of goods sold on the budgeted income statement. It will also give the dollar value of the ending materials and finished goods inventory to appear on the budgeted balance sheet.

Example 6.7 — The Johnson Company's ending inventory budget for the year ending December 31, 2001 is as follows:

	Ending Inventory Units	Unit Cost	Total
Direct materials	250 pounds (see Example 6.4)	\$ 2	\$ 500
Finished goods	100 units (see Example 6.3)	\$ 41	\$ 4,100
Total			\$ 4,600

The unit variable cost of $41 is computed as follows:

	Unit Cost	Units	Total
Direct materials	$2	3 pounds	6
Direct labor	5	5 hours	25
Variable overhead	2	5 hours[a]	10
Total variable manufacturing cost			$41

[a] Variable overhead is applied on the same basis as direct labor hours.

6.7.7 The Selling and Administrative Expense Budget

The selling and administrative expense budget lists the operating expenses incurred when selling the products and managing the business. In order to complete the budgeted income statement in contribution format, the variable selling and administrative expense per unit must be computed.

Example 6.8 — The Johnson Company's selling and administrative expense budget for the year ending December 31, 2001 is as follows.

	Quarter				
	1	2	3	4	Total
Expected sales in units	800	700	900	800	3,200
Variable selling and administrative expense per unit[a]	× $4	× $4	× $4	× $4	× $4
Budgeted variable expense	$3,200	$2,800	$3,600	$3,200	$12,800
Fixed selling and administrative expenses:					
Advertising	1,100	1,100	1,100	1,100	4,400
Insurance	2,800				2,800
Office salaries	8,500	8,500	8,500	8,500	34,000
Rent	350	350	350	350	1,400
Taxes			1,200		1,200
Total budgeted selling and administrative expenses[b]	$15,950	$12,750	$14,750	$13,150	$56,600

[a] Includes sales agents' commissions, shipping, and supplies.
[b] Paid for in the quarter incurred.

6.7.8 The Cash Budget

The cash budget is prepared to forecast financial needs. It is a tool for cash planning and control and for formulating investment strategies. Because the cash budget details the expected cash receipts and disbursements for a designated time period, it helps avoid the problem of either having idle cash on hand or suffering a cash shortage. If a cash shortage is experienced, the cash budget indicates whether the shortage is temporary or permanent and whether short-term or long-term borrowing is needed. Is a line of credit necessary? Should capital expenditures be cut back? If the cash position is very poor, the company may even go out of business because it cannot pay its bills. On the other hand, if the cash position is excessive, the company may be missing opportunities to use its cash for the greatest profit possibilities. Just as forecasts must be continuously evaluated for changing internal and external circumstances, the cash budget must also be periodically updated for changing conditions.

A cash budget for your department is similar to a cash budget you would prepare for yourself. To the beginning cash balance, you add the period's cash receipts, which gives you the total amount available to spend. You then subtract cash payments, leaving the ending cash balance. Just as you pay cash for living costs (e.g., food and shelter), a company also pays cash to operate and generate revenue (e.g., advertising). Unlike your personal cash budget, however, the company's cash budget cannot count on receiving a paycheck each period; you instead must estimate when credit sales will be collected. However, a company has more alternative sources of cash than you do. For example, a company may issue stock but a person does not.

The cash budget usually is comprised of four major sections:

1. The *receipts section* gives the beginning cash balance, cash collections from customers, and other receipts (e.g., borrowing money, selling assets). Note that cash receipts are not necessarily the same as revenue (e.g., credit sales).
2. The *disbursements section* shows all cash payments listed by purpose. Examples include cash expenses, purchase of assets, and payment of debt. The *receipts section* gives the beginning cash balance; cash collections from customers are cash payments (e.g., depreciation).
3. The *cash surplus* or *deficit section* simply shows the difference between the cash receipts section and the cash disbursements section.
4. The *financing section* provides a detailed account of the borrowings and repayments expected during the budget period.

Example 6.9 — To illustrate the cash budget we will make the following assumptions:

- The company desires to maintain a $5,000 minimum cash balance at the end of each quarter.
- All borrowing and repayment must be in multiples of $500 at an interest rate of 10% per annum. Interest is computed and paid as the principal is repaid. Borrowing takes place at the beginning of each quarter, and repayment is made at the end of each quarter.

The Johnson Company's cash budget for the year ending December 31, 2001 is shown as follows:

	Example No.	Quarter 1	2	3	4	Total
Cash balance, beginning	Given	$10,000	$9,401	$5,461	$9,106	$10,000
Receipts						
Collection from customers	6.2	54,300	57,120	66,080	64,960	242,460
Total cash available		64,300	66,521	71,541	74,066	252,460
Less: Disbursements						
Direct materials	6.4	4,549	4,560	4,860	5,113	19,082
Direct labor	6.5	19,750	18,000	22,250	20,500	80,500
Factory overhead	6.6	10,650	9,950	11,650	10,950	43,200
Selling and administrative	6.8	15,950	12,750	14,750	13,150	56,600
Machinery purchase	Given		24,300			24,300
Income tax	Given	4,000				4,000
Total disbursements		54,899	69,560	53,510	49,713	227,682
Cash surplus (deficit)		9,401	(3,039)	18,031	24,353	24,778
Financing						
Borrowing			8,500			8,500
Repayment				(8,500)		(8,500)
Interest				(425)		(425)
Total financing			8,500	(8,925)		(425)
Cash balance, ending		$9,401	$5,461	$9,106	$24,353	$24,353

6.7.9 The Budgeted Income Statement

The budgeted income statement summarizes the various component projections of revenue and expenses for the budgeting period (for control purposes, the budget can be divided into quarters or even months, depending on need).

Example 6.10 — The Johnson Company's budgeted income statement for the year ending December 31, 2001 is as follows:

	Example No.		
Sales (3,200 units @ $80)	6.2		$256,000
Less: Variable expenses			
Variable cost of goods sold			
(3,200 units @ $41)	6.7		$131,200
Variable selling and administrative	6.8	12,800	144,000
Contribution margin			112,000
Less: Fixed expenses			
Factory overhead	6.6	24,000	
Selling and administrative	6.8	43,800	67,800
Net operating income			44,200
Less: Interest expense	6.9		425
Net income before taxes			43,775
Less: Income taxes	20%		8,755
Net income			$35,020

6.7.10 The Budgeted Balance Sheet

The budgeted balance sheet alerts you to unfavorable financial conditions that you might want to avoid. It also serves as a final check on the mathematical accuracy of all the other schedules, helps you perform a variety of ratio calculations, and highlights future resources and obligations. The budgeted balance sheet is prepared by adjusting the previous year's balance sheet based on expectations for the budgeting period.

Example 6.11 — To illustrate, we will use the Johnson Company's balance sheet as of December 31, 2000.

Assets		**Liabilities and Stockholders' Equity**	
Current assets		Current liabilities	
Cash	$10,000	Accounts payable	$2,200
Accounts receivable	9,500	Income tax payable	4,000
Material inventory	474	Total current liabilities	6,200
Finished goods inventory	3,280	Stockholders' equity	
Total current assets	23,254	Common stock, no-par	70,000
Fixed assets		Retained earnings	37,054
Land	50,000		
Buildings and equipment	100,000		
Accumulated depreciation	(60,000)		
Total fixed assets	90,000		
Total assets	$113,254	Total liabilities and stockholders' equity	$113,254

The Johnson Company's budgeted balance sheet as of December 31, 2001 is shown as follows:

Assets		**Liabilities and Stockholders' Equity**	
Current assets		Current liabilities	
Cash	$24,353[a]	Accounts payable	$2,464[h]
Accounts receivable	23,040[b]	Income tax payable	8,755[i]
Material inventory	500[c]	Total current liabilities	11,219
Finished goods inventory	4,100[d]	Stockholders' equity	
Total current assets	$ 51,993	Common stock, no-par	70,000[j]
Fixed assets		Retained earnings	72,074[k]
Land	50,000[e]		
Buildings and equipment	124,300[f]		
Accumulated depreciation	(73,000)[g]		
Total fixed assets	101,300		
Total assets	$153,293	Total liabilities and stockholders' equity	$153,293

[a] From Example 6.9 (cash budget).

[b] $9,500 + $256,000 sales (from Example 6.2) – $242,460 receipts (from Example 6.9) = $23,040.

[c] From Example 6.7 (ending inventory budget).

[d] From Example 6.7 (ending inventory).
[e] No change.
[f] $100,000 + $24,300 (from Example 6.9) = $124,300.
[g] $60,000 + $13,000 (from Example 6.6) = $73,000.
[h] $2,200 + $19,346 (from Example 6.4) – $19,082 (from Example 6.4) = $2,464 (all accounts payable relate to material purchases), or 50% of fourth quarter purchase = 50% ($4,928) = 2,464.
[i] From Example 6.10 (budgeted income statement).
[j] No change.
[k] $37,054 + $35,020 (from Example 6.10) net income = $72,074.

6.8 IS THERE A SHORTCUT APPROACH TO FORMULATING THE BUDGET?

There is a shortcut approach that is widely used in formulating a budget. In the first step of this approach, a *pro forma* income statement is developed using past percentage relationships between relevant expense and cost items and sales. These percentages are then applied to forecasted sales. This is a version of the percent-of-sales method previously discussed. Second, a *pro forma* balance sheet is estimated by determining the desired level of certain balance sheet items and by making additional financing conform to those desired figures. Thus, the remaining items are estimated to make the balance sheet even.

There are two basic assumptions underlying this approach:

1. Previous financial condition is a reliable predictor of future condition.
2. The value of certain variables (e.g., cash, inventory, and accounts receivable) can be forced to take on specified *desired* values.

6.9 CAN YOU USE AN ELECTRONIC SPREADSHEET TO DEVELOP A BUDGET PLAN?

An electronic spreadsheet may be used to efficiently and quickly develop a budget (examples of available software are Lotus 1-2-3 and Microsoft Excel). Such software carries out tedious, mechanical calculations including addition, subtraction, multiplication, and division. More importantly, if a budgetary figure is changed (e.g., sales), the electronic spreadsheet will change all other budgetary amounts related to it (e.g., cost of sales, selling expenses). Hence, it is easy to make different sales projections and see what the resulting costs and profitability will be.

6.10 COMPUTER-BASED MODELS FOR FINANCIAL PLANNING AND BUDGETING

A computer-based quantitative model may be used to construct a profit planning budget and to help answer a variety of "what-if" questions. Its calculations provide a basis for choice among alternatives under conditions of uncertainty. There are primarily two approaches to modeling the corporate budgeting process: *simulation* and *optimization* (see an advanced finance text for more information on these models). There are many user-oriented modeling languages specifically designed for budgeting; among the well-known system packages are Comshare's Interactive Financial Planning System (IFPS), Venture, Encore! Plus, and MicroFCS.

6.11 CONCLUSION

Budgets should be participative, not authoritative, and the budget process should be decentralized to departmental cost centers with responsible managers. If the people operating under the budget have helped prepare it, they will be more receptive to meeting its objectives. Budgets must also be realistic if they are to be accepted by the staff and if the company's objectives are to be met. If you feel a budget is fair and you have participated in the budgetary process, you can better understand the impact of your decisions on your department, other departments, and the company as a whole. You can also determine how to best use allocated funds to strengthen your department's position.

A budget helps you direct attention from the present to the future. It assists you in anticipating problems in time to deal with them effectively, and it reveals whether you can expand. Further, the budget helps you achieve departmental goals. You have a reference point for control reporting. The budgetary process also assists you in searching for weaknesses, which you can strengthen before they become a major problem.

7 Using Variance Analysis as a Financial Tool

Variance analysis is a tool managers use to evaluate financial performance. You can carry out variance analysis for such areas of the company as sales, purchasing, and distribution. Variance analysis can also help you appraise the productivity of your employees, including sales personnel, factory workers, and delivery personnel, among others. By using variance analysis, you can spot and correct problems, as well as identify areas of success.

Locating deviations between expected and actual figures through variance analysis helps you not only to identify problems but also to determine how large the problem is, who is responsible, and whether it is controllable or not. For example, if the cost of manufacturing a product or rendering a service is higher than expected, you would want to know the reasons for the discrepancy and the severity of the situation. With the aid of variance analysis, you can then decide whether the problem lies with buying, planning, or general productivity. Or perhaps you find that actual sales fall short of expectations. You would want to know whether the sales manager can control the problem (e.g., poor salesperson effort) or not (e.g., an economic depression).

Variance analysis also spotlights positive performances. If the actual cost of a product is less than expected, for instance, you would want to know what you and your employees are doing correctly so as to capitalize on the situation. In this event, you would want to reward superior performances by employees or to know that the company is using new technological processes that successfully keep costs down.

Typically, only significant variances warrant investigation. To help you interpret variance analysis reports, understand which variances are important for your position, and remedy any problems, this chapter will cover the following topics: (1) the standards of revenue and cost; (2) the use of variance analysis to monitor performance; (3) the evaluation of sales and cost variances; (4) the use of flexible budgeting; and (5) analysis of warehouse problems and sales performance. You will learn to calculate and appraise variances so as to answer these questions:

1. Do your actual sales and expenses match your estimates? Why or why not?
2. Is the variance favorable or unfavorable? if it is favorable, who should get credit for it? If it is unfavorable, is it controllable?
3. Is the variance uncontrollable because of external factors such as a strike?
4. Are your standards too tight?
5. What trends do the variances reveal?
6. If the variance is significant (or controllable), what are the reasons and who is responsible? What corrective actions should be taken?

In an interview, a nonfinancial manager neatly summarized the virtues of variance analysis, which you are about to learn:

> Let's say production has dropped and I am below plan. I must take appropriate action to get back on course. We have a weekly production meeting where we must report on our variances to plan in a number of areas. We do this in order to maintain tight management control and also to be responsive to our changing environment. Also, this is a very important tool for planning as we each week determine the best use for our resources, budgeting our people if you will.

7.1 DEFINING A STANDARD

At the beginning of the period, you set standards such as sales quotas, standard costs (e.g., material price, wage rate), and standard volume. Let us focus on standard cost, which is the predetermined cost of manufacturing, servicing, or marketing an item during a given future period. It is based on current and projected conditions. This norm is also dependent upon quantitative and qualitative measurements (e.g., working conditions). A nonfinancial manager commented in an interview that his facilities department uses standard costs for equipment installation and relocation to determine the cost to charge to the ordering unit. For determining the projected cost of a job, you should understand the concept of and calculations for standard costs.

Variance analysis compares standard to actual costs or performances. Contingent on the cost-benefit relationship, these analyses may be as detailed as necessary. Evaluation of variances may be done yearly, quarterly, monthly, daily, or hourly, depending on how quickly you need to identify a problem. But since you do not know actual figures (e.g., hours spent) until the end of the period, variances can only be computed at that time.

There are two types of variances: material and immaterial. For material variances, you must identify the responsible departments or persons and take corrective action. Insignificant, or immaterial, variances need not be looked into further unless they recur and/or reflect potential difficulty.

One way to measure if a variance is significant is to divide the variance by the standard cost. Generally, a variance of more than 5% may be deemed material. In some cases, materiality is measured by dollar amount or by volume level. For example, the company may set a policy looking into any variance that exceeds $10,000 or 20,000 units, whichever is less. Guidelines for materiality should be set to reflect the importance of an element to the corporation's performance and decision making. For example, when a product or activity is critical to the proper functioning of the business (e.g., critical part, promotion, repairs), limits for materiality should be such that early reporting of variances is encouraged. Further, statistical techniques can be used to ascertain the significance of cost and revenue variances.

7.2 THE USEFULNESS OF VARIANCE ANALYSIS

Setting standards and analyzing variances are essential in financial analysis and decision making. Both activities tell you how you are doing and what is going wrong or right. Some specific advantages of standards and variances are that they:

- Assist in decision making because areas of inefficiency are identified.
- Help formulate sales price because you can determine what the item should cost.
- Set and evaluate objectives by comparing budgeted to actual figures.
- Establish cost control since cost inefficiencies and overruns are pinpointed.
- Highlight problem areas through the "management by exception" principle.
- Pinpoint responsibility for undesirable performance so that corrective action may be taken. Variances in product activity (cost, quantity, quality) are typically the foreman's responsibility. Variances in sales are often the responsibility of the marketing manager. Variances in profit usually relate to overall operations. Note that variances indicating strengths should be further taken advantage of.
- Facilitate communication within the organization, such as between management and supervisors, because everyone knows the standard to accomplish.
- Assist in planning by forecasting needs (e.g., cash requirements).
- Establish bid prices on contracts because desired profit goals can be determined.
- Simplify bookkeeping procedures by keeping the records at standard cost.

Although standard costing and variance analysis are exceptionally important tools, they can have their drawbacks. For example, biases in deriving standards can exist. Further, material variances can sometimes be due to out-of-date standards or poor budgetary processes rather than actual performances. Therefore, you should use these tools with caution.

7.3 SETTING STANDARDS

There are four types of standards:

- **Basic** — These are not changed from period to period. They form the basis with which later period performance is compared. No consideration, however, is given to a change in the environment, which is an unrealistic basis.
- **Maximum efficiency** — These are perfect standards, which assume ideal, optimal conditions. Realistically, certain inefficiencies will occur.
- **Currently attainable** — These are based on efficient activity. They are possible goals but difficult to achieve. Normal occurrences, such as anticipated machinery failure, are considered in the calculation at currently attainable standards.
- **Expected** — These are expected figures, which should come very close to actual figures.

You should base standards on the situation being appraised. Here are some examples of how to match standards with particular situations:

Situation	Standard
Cost reduction	Tight
Pricing policy	Realistic
High-quality good	Perfection

7.4 DETERMINING AND EVALUATING SALES VARIANCES

Actual and budgeted sales figures should be compared by territory, salesperson, and product to see if expectations are being realized.

To evaluate sales performance, you should focus on *profitable* sales volume. Do not be misled by *high* sales volume; high sales volume does not automatically mean high profits because the high costs associated with products need to be subtracted. Thus, reports by salespersons should show not only revenue derived but also gross profit earned. Although gross profit may not necessarily translate into net profit because of high selling and advertising expenses, it is a legitimate measure of performance. Overall, the report will reveal where to put sales effort.

When you appraise changes in gross profit, you have to recognize that these changes may be caused by changes in sales mix, selling price, volume, or returns. Also, a complete analysis of sales volume should include consideration of budgets, standards, sales plans, industry comparisons, and manufacturing costs.

Sales variances are computed to gauge the performance of the marketing function.

Example 7.1 — Western Corporation's budgeted sales for 2001 were:

Product A: 10,000 units at $6.00 per unit	$ 60,000
Product B: 30,000 units at $8.00 per unit	240,000
Expected sales revenue	$300,000

Actual sales for the year were:

Product A: 8,000 units at $6.20 per unit	$ 49,600
Product B: 33,000 units at $7.70 per unit	254,100
Actual sales revenue	$303,700

There is a favorable sales variance of $3,700, consisting of the sales price variance and sales volume variance. The sales price variance equals:

(Actual selling price – Budgeted selling price) × Actual units sold

Product A ($6.20 – $6.00) × 8,000	$1,699	Favorable
Product B ($7.70 – $8.00) × 33,000	9,900	Unfavorable
Sales price variance	$8,300	Unfavorable

The sales volume variance equals

(Actual quantity – Budgeted quantity) × Budgeted selling price

Product A (8,000 – 10,000) × $6.00	$12,000	Unfavorable
Product B (33,000 – 30,000) × $8.00	24,000	Favorable
Sales volume variance	$12,000	Favorable

By looking at the trend in sales, you may identify unfavorable trends requiring a reduction of the sales effort or a change in product.

You have to evaluate the meaning of sales variances to your department. For example, an unfavorable sales volume variance may arise from poor marketing procedures. An unfavorable total sales variance may signal a problem with the marketing manager because he or she has control over sales, advertising, and often pricing. Unfavorable sales situations may also be caused by a lack in quality control, substitution of poorer quality components because of deficient purchasing, or deficient product design based on poor engineering.

You should also compare your divisional sales figures with the sales figures of competing companies' divisions and with industry averages. Sometimes an unfavorable sales volume variance is caused by the price cuts of competing companies. And, if unfavorable volume variance is coupled with favorable price variance, you may have lost sales by raising your prices.

7.5 COST VARIANCES

When a product is made or a service is performed, you have to determine these three cost measures:

- Actual cost equals actual price times actual quantity, where actual quantity equals actual quantity per unit of work times actual units of work produced.
- Standard cost equals standard price times standard quantity, where standard quantity equals standard quantity per unit of work times actual units of work produced.
- Total cost variance equals actual cost less standard cost.

Total cost variance has these two elements: price variance and quantity variance. Price (rate, spending) variance is calculated as follows:

$$(\text{Standard price} - \text{Actual price}) \times \text{Actual quantity}$$

Quantity (usage, efficiency) variance is formulated in the following manner:

$$(\text{Standard quantity} - \text{Actual quantity}) \times \text{Standard price}$$

These figures are computed for both material and labor. Figure 7.1 depicts the variance analysis for total cost.

7.6 MATERIALS VARIANCES

The materials price variance allows you to evaluate the activity of the purchasing department and to see the impact of raw materials' cost changes on profitability, while the materials quantity variance helps you to judge the performance of the production supervisor. Example 7.2 calculates the three types of materials variances you need to know.

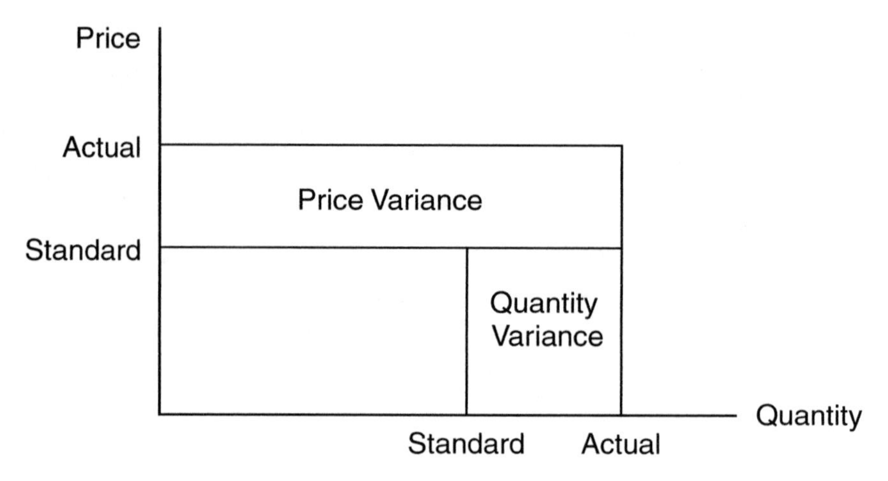

FIGURE 7.1 Variance analysis of total cost.

Example 7.2 — The standard cost of one unit of output (product or service) was \$15, or three pieces at \$5 per piece. During the period, 8,000 units were made. The actual cost was \$14 per unit, or two pieces at \$7 per piece.

Total Materials Variance

Standard quantity × Standard price (24,000 × \$5)	\$120,000
Actual quantity × Actual price (16,000 × \$7)	112,000
	\$8,000 F

Materials Price Variance

(Standard price – Actual price) × Actual quantity	
(\$5 – \$7) × 16,000	\$32,000 U

Materials Quantity Variance

(Standard quantity – Actual quantity) × Standard price	
(24,000 – 16,000) × \$5	\$40,000 F

Unfavorable materials price variances may be caused by one or more of the following factors: inaccurate standard prices, failure to take a discount on quantity purchases, failure to shop for bargains, inflationary cost increases, scarcity in raw material supplies resulting in higher prices, and purchasing department inefficiencies.

Unfavorable materials quantity variances result from poorly trained workers, improperly adjusted machines, use of an improper production method, outright waste on the production line, and/or use of a lower-grade material, purchased to save money. Once the reasons for unfavorable materials variances have been determined, you can then locate the responsible departments or key personnel:

Reason	Responsible Party
Overstated price paid	Purchasing
Failure to detect defective goods	Receiving
Inefficient labor or poor supervision	Foreman
Poor mix in material	Production manager
Rush delivery of materials	Traffic
Unfavorable quantity variance	Foreman
Unexpected change in production volume	Sales manager

To correct an unfavorable materials price variance, you can increase selling price, substitute cheaper materials, change a production method or specification, or engage in a cost reduction program. To achieve some of these ends, you should examine the variability in raw material costs, look at price instability in trade publications, and emphasize vertical integration to reduce the price and supply risk of raw materials.

Lastly, remember that you cannot control materials price variances when higher prices are caused by inflation or shortage situations or when rush orders are required by the customer, who will bear the ultimate cost increase.

7.7 LABOR VARIANCES

The standard labor rate is based on the contracted hourly wage rate. Where salary rates are set by union contract, the labor rate variance will usually be minimal. Labor efficiency standards are typically estimated by engineers based on an analysis of the production operation.

Labor variances are determined in a manner similar to that of materials variances, as illustrated in Example 7.3.

Example 7.3 — The standard cost of labor is four hours times $9 per hour, or $36 per unit. During the period, 7,000 units were produced. The actual cost is six hours times $8 per hour, or $48 per unit.

Total Labor Variance

Standard quantity × Standard price (28,000 × $9)	$252,000
Actual quantity × Actual price (42,000 × $8)	336,000
	$84,000 U

Labor Rate Variance

(Standard price – Actual price) × Actual quantity	
($9 – $8) × 42,000	$42,000 F

Labor Efficiency Variance

(Standard quantity – Actual quantity) × Standard price	
(28,000 – 42,000) × $9	$126,000 U

Possible causes of *unfavorable* labor rate variance include the following: (1) an increase in wages; (2) poor scheduling or production resulting in overtime work; and (3) use of workers commanding higher hourly rates than expected.

Possible reasons for labor rate variances can be matched with responsible parties, as follows:

Reason	Responsible Party
Use of overpaid or an excessive number of workers	Production manager or union contract
Poor job descriptions	Personnel
Overtime	Production planning

In the case of a shortage of skilled workers, it may be impossible to avoid an unfavorable labor price variance.

Labor efficiency (quantity) variances can be attributed to one or more of the following:

- Poor supervision
- Use of unskilled workers, who are paid lower rates, or the wrong mixture of labor for a given job
- Use of poor quality machinery
- Improperly trained workers
- Poor quality of materials, requiring more labor time in processing
- Machine breakdowns
- Employee unrest
- Production delays due to power failure

The reasons and responsible parties for unfavorable labor efficiency variances can then be matched in these ways:

Reason	Responsible Party
Inadequate supervision	Foreman
Improper functioning of equipment	Maintenance
Insufficient material supply or poor quality	Purchasing

7.8 OVERHEAD VARIANCES

The overhead variance is composed of the controllable and volume variances. Relevant computations follow.

Overhead control variance = Actual overhead vs. Standard overhead (standard hours $\times$ standard overhead rate).

Controllable variance = Actual overhead vs. Budget adjusted to standard hours.

Note: Budget adjusted to standard hours equals fixed overhead plus variable overhead (standard hours times standard variable overhead rate).

Volume variance = Standard overhead vs. Budget adjusted to standard hours.

Example 7.4 uses the equations to illustrate a hypothetical situation.

Example 7.4 — The following data are provided:

Budgeted overhead (includes fixed overhead of $7,500 and variable overhead of $10,000)	$17,500
Budgeted hours	10,000
Actual overhead	8,000
Actual units produced	800
Standard hours per unit of production	5

Preliminary calculations are as follows:

Budgeted fixed overhead ($7,500/10,000 hrs)	0.75
Budgeted variable overhead ($10,000/10,000 hrs)	1.00
Total budgeted overhead ($17,500/10,000 hrs)	$1.75
Standard hours (800 units × 5 hrs per unit)	$4,000

Total Overhead Variance

Actual overhead		$8,000
Standard overhead		
Standard hours	4,000 hr	
Standard overhead rate	× $1.75	
	7,000	7,000
		$1,000U

Controllable Variance

Actual overhead		$8,000
Budget adjusted to standard hours:		
Fixed overhead	$7,500	
Variable overhead (Standard hours × Standard Variable overhead rate: 4,000 × $1)	4,000	
	11,500	11,500
		$3,500F

Volume Variance

Standard overhead	7,000
Budget adjusted to standard hours	11,500
	$4,500U

Because unfavorable variances exist, you would want to locate the responsible parties. The controllable variance is the responsibility of the foreman, since he or she influences actual overhead incurred. The volume variance is the responsibility of management executives and production managers, since they are involved with plant utilization.

On the one hand, variable overhead variance information is helpful in arriving at output level and output mix decisions. It also assists in appraising decisions regarding variable inputs. On the other hand, fixed overhead variance data provide information regarding decision-making astuteness in relation to the purchase of some combination of fixed plant size and variable production inputs.

The possible reasons for recurring unfavorable overhead volume variances include the following:

- The purchase of the wrong size plant
- Improper scheduling
- Insufficient orders
- Shortages in material
- Machinery failure
- Long operating time
- Inadequately trained workers

When idle capacity exists, this may indicate long-term operations' planning problems.

7.9 THE USE OF FLEXIBLE BUDGETS IN PERFORMANCE REPORTS

The *static* (fixed) budget is geared to only one level of activity and thus has problems in cost control because it does not distinguish between fixed costs and variable costs.

The *flexible* budget is geared toward a range of activities rather than a single level of activity. A flexible budget employs budgeted figures at different capacity levels. It allows you to choose the best expected (normal) capacity level (100%) and to assign pessimistic (80%), optimistic (110%), and full (150%) capacity levels. You can then see how your department is performing at varying capacity levels. Fixed costs remain constant as long as you operate below full capacity.

Flexible budgeting distinguishes between fixed and variable costs, thus allowing for a budget that can be automatically adjusted (via changes in variable cost totals) to the particular level of activity actually attained. Thus, variances between actual costs and budgeted costs are adjusted for volume ups and downs before differences due to price and quantity factors are computed. Note that flexible budgets are primarily used to accurately measure performance because they compare actual costs for a given output with the budgeted costs for the *same level of output*.

Table 7.1 presents an example of a flexible budget. Example 7.5 illustrates the differences between static and flexible budgets.

TABLE 7.1
Flexible Budget

Cost	Variable Cost	Fixed Cost	Budget (50,000 hrs.)	Actual (60,000 hrs.)	Budget (60,000 hrs.)	Spending Variance
Indirect material	$.80	$6,000	$46,000	$56,000	$54,000	$2,000(U)
Indirect labor	4.20	24,000	234,000	270,000	276,000	6,000(F)
Supervision		70,000	70,000	74,000	70,000	4,000(U)
Depreciation		40,000	40,000	32,000	40,000	8,000(F)
Power and light	.30	2,000	17,000	21,000	20,000	1,000(U)
Maintenance	4.70	58,000	293,000	287,000	340,000	53,000(F)
Total	$10.00	$200,000	$700,000	$740,000	$820,000	$60,000(F)

Example 7.5 — Assume that the Fabricating Department of Company X is budgeted to produce 6,000 units during June. Assume further that only 5,800 units were produced. The direct labor and variable overhead budget for the month of June follows:

Budgeted production	6,000 units
Actual production	5,800 units
Direct labor	$39,000
Variable overhead costs	
Indirect labor	6,000
Supplies	900
Repairs	300
	$46,200

If a static budget is used, the performance report for Company X's Fabricating Department will appear as follows:

	Budget	Actual	Variance (U or F)
Production in units	6,000	5,800	200U
Direct labor	$39,000	$38,500	$500F
Variable overhead costs			
Indirect labor	6,000	5,950	50F
Supplies	900	870	30F
Repairs	300	295	5F
	$46,200	$45,615	$585F

These cost variances are useless: They compare oranges with apples. That is, the budget costs are based on an activity level of 6,000 units, whereas the actual costs were incurred at an activity level below this figure (5,800 units). From a control standpoint, it makes no sense to compare costs at one activity level with costs at a different one. Such comparisons make a production manager look good as long as the actual production is less than the budgeted production. For our current examples, we should compare costs at the actual production of 5,800 units.

Budgeted production	6,000 units
Actual production	5,800 units

		Budget (5,800 units)	Actual (5,800 units)	Variance (U or F)
Direct labor	$6.50 per unit	$37,700	$38,500	$800U
Variable overhead				
Indirect labor	1.00	5,800	5,950	150U
Supplies	.15	870	870	0
Repairs	.05	290	295	5U
	$7.70	$44,660	$45,615	$955U

All cost variances are unfavorable, as compared to the favorable cost variance on the performance report using the static budget approach.

7.10 STANDARDS AND VARIANCES IN MARKETING

Prior to setting a marketing standard in a given trade territory, you should examine prior, current, and forecasted conditions. You should also make adjustments for standards that vary with geographical location. In formulating standard costs for the transportation function, minimum cost traffic routes should be selected based on the distribution pattern.

Standards for advertising cost in particular territories will vary depending upon the types of advertising media needed, which are in turn based on the type of customers the advertising is intended to reach as well as the nature of the competition.

Some direct selling costs can be standardized, such as product presentations for which a standard time per sales call can be established. Direct selling expenses should be related to distance traveled, frequency of calls made, etc. If sales commissions are based on sales generated, standards can be based on a percentage of net sales.

Time and motion studies are usually a better way of establishing sales standards than prior performance, since past strategies may have been inefficient.

Of the numerous marketing standards and variances, we will focus on, for the purpose of illustration, those associated with sales.

7.10.1 SALES STANDARDS

Sales standards may be set in terms of effort, accomplishments, or the relationship between effort and accomplishment. For example, a salesperson may be expected to make 100 calls per month. By making 100 calls, the salesperson meets a standard. He or she may be expected to obtain 60 orders or $300,000 in business based on the 100 calls. By doing so, a different standard is accomplished. You need to set sales standards to control sales activities, to reward merit, and to encourage sales effort.

7.10.2 Analyzing Salesperson Variances

You should appraise sales force effectiveness within a territory, including time spent and expenses incurred. Examples 7.6 and 7.7 provide you with the data and computations for analyzing variances in two hypothetical situations.

Example 7.6 — Sales data for your department follows.

Standard cost	$240,000
Standard salesperson days	2,000
Standard rate per salesperson day	120
Actual cost	238,000
Actual salesperson days	1,700
Actual rate per salesperson day	140

Total Cost Variance

Actual cost	$238,000
Standard cost	240,000
	$ 2,000F

Note: Cost variances may be determined by territory, product, or personnel. The control variance is broken down into salesperson days and salesperson costs.

Variance in Salesperson Days

(Actual days – Standard days) × Standard rate per day	
(1,700 – 2,000) × $120	$36,000 F

The variance is favorable because the territory was handled in fewer days than expected.

Variance in Salesperson Costs

(Actual rate – Standard rate) × Actual days	
($140 – $120) × 1,700	$34,000 U

An unfavorable variance results because the actual rate per day is greater than the expected rate per day.

Example 7.7 — A salesperson called on 55 customers and sold each an average of $2,800 of merchandise. The standard of calls is 50 and the standard sale is $2,400. Variance analysis of calls and sales follows.

Total Variance

Actual calls × Actual sales (55 × $2,800)	$154,000
Standard calls × Standard sales (50 × $2,400)	120,000
	$34,000 F

The $34,000 favorable variance can be broken down as follows.

Variance in Calls

Actual calls – Standard calls) × Standard sale	
(55 – 50) × $2,400	$12,000 F

Variance in Sales

(Actual sale – Standard sale) × Standard calls	
($2,800 – $2,400) × 50	$20,000 F

Joint Variance

(Actual calls – Standard calls) × (Actual sale – Standard sale)	
(55 – 50) × ($2,800 – $2,400)	$2,000F

You should encourage your sales staff to push products with the highest *profitability* rather than the highest selling price or least sales resistance. Additional performance measures of sales force effectiveness include meeting sales quotas, the number of orders from current and new customers, profitability per order, and the relationship between salesperson costs and revenue obtained.

Avoid paying your salespeople commissions on gross sales because you may encourage them to sell to anyone, resulting in huge sales returns and allowances. Instead, the commission should be tied to net sales.

7.11 VARIANCES IN WAREHOUSING COSTS

Variance in warehousing costs can be calculated by looking at the cost per unit to store the merchandise and the number of orders anticipated. Example 7.8 provides the data and calculations of the warehousing cost variance for a hypothetical situation.

Example 7.8 — The following information applies to a product:

Standard cost	$12,100
Standard orders	5,500
Standard unit cost	2.20
Actual cost	14,030
Actual orders	6,100
Actual unit cost	$ 2.30

Total Warehousing Cost Variance

Actual cost	$14,030
Standard cost	12,100
	$1,930 U

The total variance is segregated into the variance in orders and variance in cost, as follows.

Variance in Orders

(Actual orders vs. Standard orders) × Standard unit cost	
(6,100 vs. 5,500) × \$2.20	\$1,320 U

Variance in Cost

(Actual cost per unit vs. Standard cost per unit) × Actual orders	
(\$2.30 vs. \$2.20) × 6,100	\$610 U

7.12 CONCLUSION

Variance analysis is essential for the appraisal of all aspects of the business, including manufacturing, marketing, and service. Variances are red (or green) flags to watch for and become monitoring and control devices. Unfavorable variances must be examined to ascertain whether they are controllable by you or uncontrollable because they solely relate to external factors. When a variance is controllable, immediate corrective action must be taken to handle the problem. If a variance is favorable, an examination should still be made so that corporate policy may include the positive aspects. Further, the party responsible for a favorable variance should be recognized and rewarded.

By looking at variances, you make your department more efficient and less costly to run, with the quality of service or product at a constant level. You should also determine an acceptable range of variances. Even a small variance can mean a large savings, if favorable, or a large loss, if unfavorable. By analyzing variances, you are using a tool to protect savings and to cut inefficiencies and losses.

Part II

Critical Asset Management Issues

8 Working Capital and Cash Management

The ability to manage working capital will improve return and minimize the risk of running short of cash. There are various ways of managing working capital to achieve success. These methods will be discussed in this chapter.

8.1 WORKING CAPITAL

Briefly, working capital equals total current assets — the most liquid assets of the company. Example 8.1 illustrates how to calculate working capital.

Example 8.1 — The following assets exist: $10,000 in cash, $30,000 in accounts receivable, $42,000 in inventory, $90,000 in machinery, $36,000 in long-term investments, and $4,000 in patents.

The working capital consists of the total current assets:

Cash	$10,000
Accounts receivable	30,000
Inventory	42,000
Total working capital	$82,000

The liquidity of the current assets will influence the terms and availability of short-term credit. The more liquid your business is, generally the easier it is for you to obtain short-term credit at favorable terms. Short-term credit, in turn, affects the amount of cash held.

Your working capital needs will partly depend on how long it takes your company to turn raw materials into sold products. The longer it takes to acquire or produce goods, the greater the working capital needed.

8.2 FINANCING ASSETS

When you need funds to buy seasonal inventory, you should finance with short-term debt because the sale of the inventory will generate funds in the near term to pay off the obligation. The short-term debt provides flexibility to meet seasonal needs within your ability to repay the loan. Long-term assets should be financed with long-term debt because the assets last longer and the financing can be spread over a longer time period. These two strategies are called *hedging*; that is, you are financing assets with liabilities of similar maturity so the proceeds and return from the assets may be used to pay off the debt, resulting in less financing risk.

8.3 MANAGING CASH PROPERLY

Sound cash management requires you to know how much cash is needed, how much cash you have, and where that cash is. This knowledge is particularly crucial during inflationary periods.

Before you plan a cash management strategy, you need to take care of some basic cash-handling needs. For example, make certain that cash receipts are deposited promptly. Also, make sure that cash is not unnecessarily tied up in other accounts such as advances or loans to employees and insurance deposits.

Next, you need to determine your cash management objective. You want to invest excess cash for a return but have adequate liquidity to meet your needs. Thus, you have to plan when to have excess funds available for investment and when money needs to be borrowed. But some available cash is needed for unexpected demands or for customer collection problems.

What determines how much cash you should hold? The key factors are cash management policies, current liquidity position, liquidity risk preferences, maturity dates of debt, ability to borrow, forecasted short- and long-term cash flow, and the probabilities of different cash flows under varying circumstances.

You do not want an excessive cash balance because no return can be earned on it. The minimum cash to hold is the greater of (1) compensating balances (a deposit held by a bank to compensate for providing services) or (2) precautionary balances (money held for emergency purposes) plus transaction balances (money to cover checks outstanding).

You can get away with a smaller cash balance when cash receipts and cash disbursements are highly synchronized and predictable. You have to accurately forecast the amount of cash needed, its source, and its destination. These data are needed on both a short- and a long-term basis. Forecasting assists in properly timing financing, debt repayment, and the amount to be transferred between accounts.

If you have many bank accounts, you may be able to guard against accumulating excessive balances. You need less cash on hand if you can borrow quickly from a bank, such as under a *line of credit agreement*, which allows you to borrow instantly up to a specified maximum amount. *Warning*: Cash in some bank accounts may not be available for investment. For example, when a bank tends your money, it may require you to retain funds on hand as collateral. This deposit is referred to as a *compensating balance*, which in effect represents *restricted cash*.

Cash management also requires knowing the amount of funds available for investment and the length of time they can be invested. You can invest funds in the following ways:

- Time deposits, including savings accounts earning daily interest, long-term savings accounts, and certificates of deposit (short-term, negotiable time deposits with banks, issued at face value and providing for interest at maturity).
- Money market funds, which are managed portfolios of short-term, high-grade debt instruments such as Treasury bills and commercial paper. Demand deposits that pay interest.

- Banker's acceptances, which are short-term (up to six months), noninterest-bearing notes up to $1 million issued at a discount. Acceptances are mostly traded internationally and enjoy their preferred credit status because of their "acceptance" (guarantee) for future payment by commercial banks. Banker's acceptances are time drafts drawn on and accepted by a banking institution, which in effect substitutes its credit for that of the importer or holder of merchandise.
- Commercial paper, which is a short-term, unsecured debt instrument issued by financially strong companies (see Chapter 16 for a more detailed discussion).
- Short-term tax exempts issued by states and local government agencies in denominations up to $1 million. The maturity typically ranges from 6 to 12 months. A good secondary market exists for resale. Interest is paid at maturity.
- Repurchase agreements issued by dealers in U.S. government securities. The maturity ranges from 1 day to 3 months. They come in typical amounts of $500,000 or more. No secondary market exists. They are legal contracts between a borrower (security seller) and lender (security buyer). The borrower will repurchase at the contract price plus an interest charge.
- Treasury bills, which are backed by the U.S. government. They have a maturity typically ranging from 90 days to 1 year. They come in denominations of up to $1 million and are highly liquid. Treasury bills are issued at a discount, which means that interest represents the difference between the price paid and the face value of the instrument at the maturity date. Ninety- and 180-day bills are auctioned weekly, whereas 270- and 360-day bills are auctioned monthly. Treasury bills are exempt from state and local income tax.

Example 8.2 demonstrates how to calculate the annual return of a Treasury bill.

Example 8.2 — A $10,000, three-month Treasury bill is purchased for $9,800. The return for three months is

$$\frac{\text{Return}}{\text{Investment}} = \frac{\$10{,}000 - \$9{,}800}{\$9{,}800} = \frac{\$200}{\$9{,}800} = 2\%$$

On an annual basis, the return is

$$2\% \times 12/3 = 8\% \text{ (annual return)}$$

By holding marketable securities, you can protect against cash shortages. If your business is seasonal, you may buy marketable securities when excess funds exist and then sell the securities when there is a cash deficit. You may also invest in marketable securities when funds are being held temporarily in anticipation of short-term capital expansion. In deciding on an investment portfolio, you should consider return rate, default risk, marketability, and maturity date.

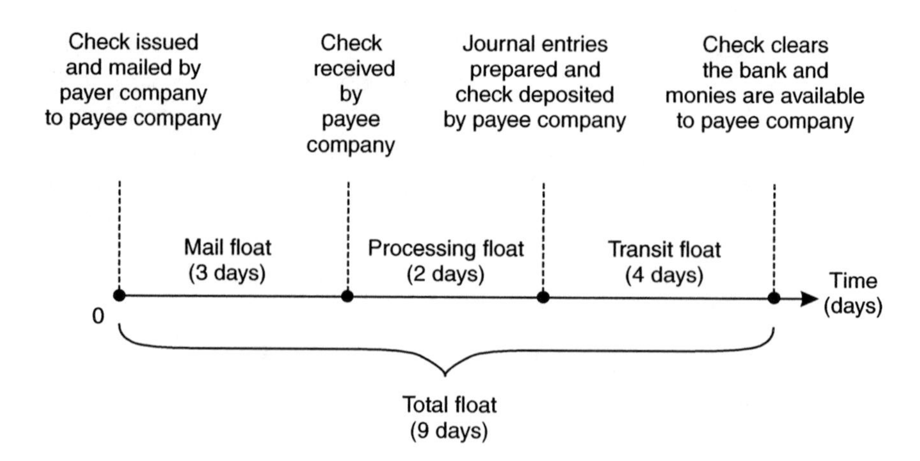

FIGURE 8.1 Total float time.

In deciding on a cash management system, you should consider its associated costs compared to the return earned from implementing the system. The costs include bank charges, your time, and office employee salaries. You can make use of your personal computer in making transactions with the computers of banks and money market funds. Computer systems are also useful for buying and selling securities in the money market.

The *thrust* of cash management is to accelerate cash receipts and delay cash payments. Analyze each bank account as to its type, balance, and cost so that return is maximized.

8.4 GETTING MONEY FASTER

To accelerate cash inflow, you must: (1) know the bank's policy regarding fund availability; (2) know the source and location of cash receipts; and (3) devise procedures for quick deposit of checks received and for quick transfer of receipts in outlying accounts into the main corporate account.

Before you draw up plans for quick deposits and transfers, you need to understand the check processing delays. Delays include: (1) mail float–the time a check moves from a debtor to a creditor; (2) processing float–the time it takes for a creditor to deposit the check after receipt; and (3) deposit collection (transit) float–the time for a check to clear. Figure 8.1 illustrates the float resulting from a check issued and mailed by the payer company to the payee company. Figure 8.2 illustrates a cash collection system for a hypothetical business.

You can minimize mail float by making sure that the collection center is close to the customer. Local banks should be selected to hasten the receipt of funds for subsequent transfer to the central corporate account. Alternatively, strategic post office lockboxes may be used for customer remissions. The local bank collects from these boxes periodically during the day and deposits the funds in the corporate account. The bank also provides a computer listing of payments received by account

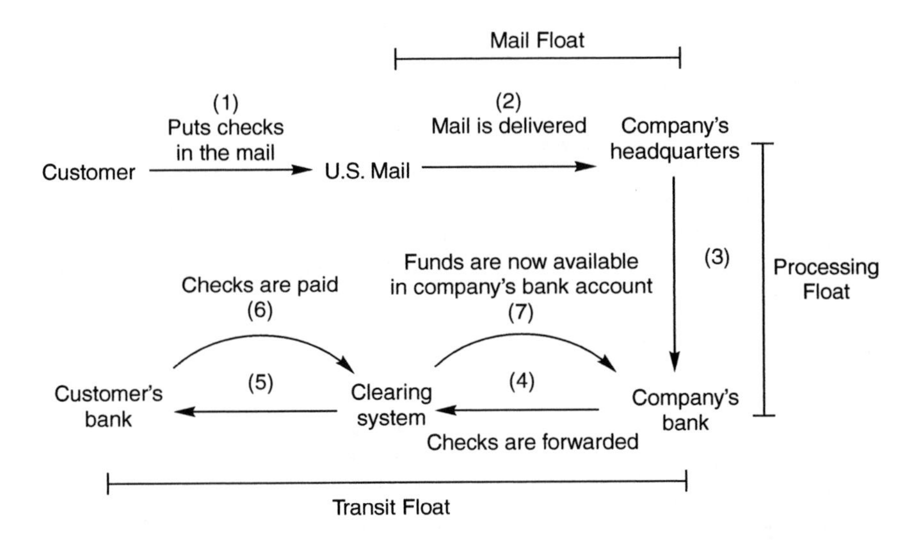

FIGURE 8.2 Cash collection system.

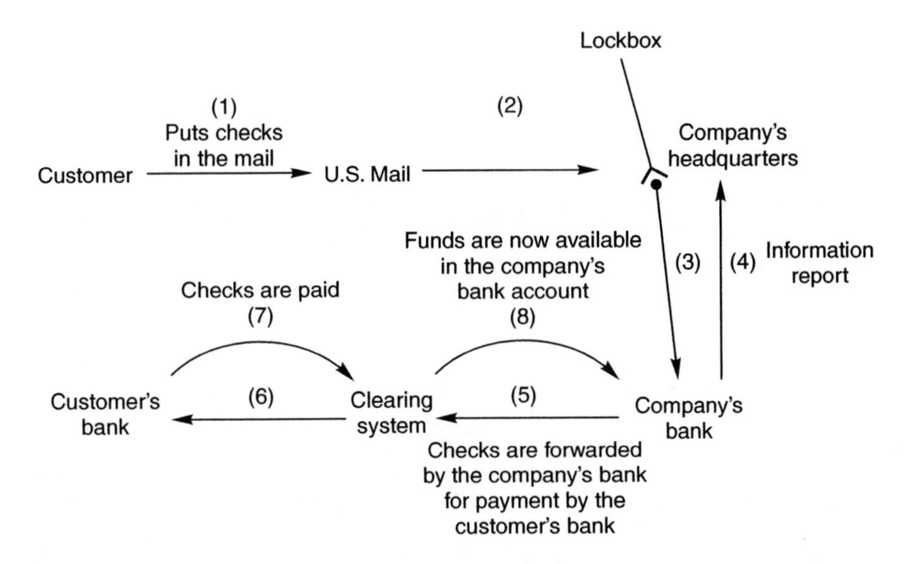

FIGURE 8.3 Lockbox system.

and a daily total. Because the lockbox system has a significant per-item cost, it is most effective with low-volume, high-dollar collections. But the system is becoming increasingly more available to firms with high-volume, low-dollar receipts because technological advances (such as machine-readable documents) are lowering the per-item cost of lockboxes.

Example 8.3 calculates, for a hypothetical situation, the amount of money that can be tied up by mailing delays.

Example 8.3 — You obtain average cash receipts of $300,000 per day. It typically takes 6 days from the time a check is mailed to its availability for use. The amount tied up by the delay is

$$6 \text{ days} \times \$300{,}000 = \$1{,}800{,}000$$

Before implementing a lockbox system, you need to undertake a cost-benefit analysis considering the average dollar amount of checks received, the costs saved by having lockboxes, the decline in mailing time per check, and the processing cost. Figure 8.3 depicts a simple lockbox system.

Example 8.4 shows you how to calculate the maximum monthly payment for a lockbox.

Example 8.4 — It takes about 8 days to receive and deposit payments from customers. Thus, a lockbox system is being considered. It is anticipated that the system will reduce the float time to 5 days. Average daily collections are $400,000. The return rate is 11%.

The reduction in outstanding cash balances occurring if the lockbox system is initiated is

$$3 \text{ days} \times \$400{,}000 = \$1{,}200{,}000$$

The return that could be earned on these funds is

$$\$1{,}200{,}000 \times .11 = \$132{,}000$$

The maximum monthly charge that should be paid for the lockbox arrangement is therefore

$$\frac{\$132{,}000}{12} = \$11{,}000$$

Example 8.5 illustrates the financial advantage of a lockbox system.

Example 8.5 — You are considering the use of a lockbox system costing $130,000 per year. Daily collections average $360,000. The lockbox arrangement will reduce the float period by 3 days. The rate of return is 14%.

The cost-benefit analysis is shown below:

Return on early collection of cash (.14 × 3 × $360,000)	$ 151,200
Less: Cost	(130,000)
Advantage of lockbox	$ 21,200

You may also have to determine whether or not it would be financially beneficial to split a geographic collection region into a number of pans. Example 8.6 demonstrates how geographical splitting can be advantageous.

Example 8.6 — You have an agreement with Harris Bank, which handles $4 million in collections a day and requires a compensating balance of $600,000. You are contemplating canceling the agreement and dividing the eastern region so that two other banks will handle the business instead. Bank X will handle $1 million a day in collections, requiring a compensating balance of $300,000, and Bank Y will handle the other $3 million a day, requiring a compensating balance of $500,000. It is expected that collections will be hastened by 1 day if the eastern region is divided. The rate of return is 12%.

The new arrangement is financially beneficial, as indicated below.

Acceleration in cash receipts ($4 million × 1/4)	$1,000,000
Less: Additional compensating balance ($800,000 – $600,000)	(200,000)
Increased cash flow	$ 800,000
Times: Rate of return	× .12
Net annual savings	$96,000

Concentration banking should also be considered where funds are collected by several local banks and transferred to a main *concentration* account in another bank. The transfer of funds may be accomplished through the use of depository transfer checks (DTCs) or wire transfers. In the DTC arrangement, you make a resolution statement with the bank, whereby signatureless checks are deposited. As the initial banks collect the funds, information is immediately transferred to the concentration bank, which then issues a DTC to collect the outlying funds. The funds may be available the same day. With wire transfers (e.g., bank wire, federal reserve wire), funds are moved immediately between banks. This eliminates transit float in that only "good funds" are transferred.

Other ways to accelerate cash receipt deposits include the following:

- Arrange for preauthorized checks where the company writes the checks for its customers to be charged against their demand deposit accounts.
- Send customers preaddressed, stamped envelopes.
- Require deposits on large or custom orders or progress billings as the work progresses.
- Charge interest on accounts receivable after a certain amount of time.
- Encourage postdated checks from customers.
- Have cash-on-delivery terms.

In accelerating cash receipts, we may want to determine the return on the cash balance.

Example 8.7 — Your company's weekly average cash balances follow:

Week	Average Cash Balance
1	$14,000
2	16,000
3	11,000
4	15,000
Total	$56,000

The monthly average cash balance is

$$\frac{56{,}000}{4} = \$14{,}000$$

Assuming an annual interest rate of 12%, the monthly return on the average cash balance is

$$\$14{,}000 \times 1\% = \$140$$

By accelerating remissions, freed cash can be invested in marketable securities for a return or used to pay off short-term debt.

8.5 DELAYING CASH PAYMENTS

You should try to keep money longer to earn greater interest! Never pay bills, for example, before their due dates. Other ways to delay cash payments include the following:

1. Make partial payments.
2. Use drafts to pay bills because drafts are not due on demand. When a bank receives a draft it must return the draft to you for acceptance before payment. When you accept the draft, you then deposit the funds with the bank; therefore, you can maintain a smaller average checking balance.
3. Mail checks from post offices with limited service or from locations where the mail must go through several handling points, lengthening the payment period.
4. Draw checks on remote banks or establish cash disbursement centers in remote locations so that the payment period is lengthened. For example, you can pay someone in New York with a check drawn on a California bank.
5. Use credit cards and charge accounts to lengthen the time between your purchase of goods and the payment date.
6. Delay the frequency of payments to employees (e.g., expense account reimbursements, payrolls).
7. Disburse sales commissions when the receivables are collected rather than when the sales are made.
8. Maintain zero balance accounts where zero balances are established for all of the company's disbursing units. These accounts are in the same concentration bank. Checks are drawn against these accounts, with the balance in each account never exceeding $0. Divisional disbursing authority is thus maintained at the local level of management. The advantages of zero balance accounts are better control over cash payments, reduction in excess cash balances held in regional banks, and a possible increase in disbursing float.

You can control cash payments by centralizing the payment operation so that obligations are satisfied at optimum times. Centralization will also aid in predicting the disbursement float.

You can minimize cash balances by using probabilities of the expected time checks will clear. Deposits, for example, may be made to a payroll checking account based on the anticipated time of check clearance. Examples 8.8 and 8.9 illustrate this principle.

Example 8.8 — You write checks averaging \$50,000 per day; each check takes 3 days to clear. You will have a checkbook balance \$150,000 less than the bank's records.

Example 8.9 — Every 2 weeks, you make out checks that average \$600,000 and take 3 days to clear. How much money can you save annually if you delay transfer of funds from an interest-bearing account that pays 0.0384% per day (annual rate of 14%) for those 3 days?

The interest for three days is

$$\$600{,}000 \times (.000384 \times 3) = \$691$$

The number of 2-week periods in a year is

$$\frac{52 \text{ weeks}}{2 \text{ weeks}} = 26$$

The savings per year is

$$\$691 \times 26 = \$17{,}966$$

Although not a delay of cash payment, you may reduce cash outflow by the early repayment of a loan, thus avoiding some interest.

A cash management system that incorporates the suggestions in the last two sections is shown in Figure 8.4.

Finally, in planning cash flows, make sure to consider that some types are more uncertain than others. For example, a major uncertain cash flow is cash sales. A major known cash flow is payroll payments. Make adjustments accordingly.

Acceleration of Cash Receipts	Delay of Cash Payments
Lock box system	Pay by draft
Concentration banking	Requisition more frequently
Preauthorized checks	Disbursing float
Preaddressed stamped envelopes	Make partial payments
Obtain deposits on large orders	Use charge accounts
Charge interest on overdue receivables	Delay frequency of paying employees

FIGURE 8.4 Cash management system.

8.6 OPPORTUNITY COST OF FOREGOING A CASH DISCOUNT

Many companies establish credit terms that authorize cash discounts in exchange for early payment of the amount purchased. If you take advantage of the cash discount, you will reduce the purchase cost.

An *opportunity cost* is the net revenue you lose by rejecting an alternative action. You should typically take advantage of a discount offered by a creditor because of the high opportunity cost. If you are short of funds to pay the supplier early, you should borrow the money when the interest rate on the loan is below the annual rate of the discount. For example, if the terms of sale are 2/10, net/30, you have 30 days to pay the bill but will get a 2% discount if you pay in 10 days.

You can use the following formula to compute the opportunity cost percentage on an annual basis:

$$\text{Opportunity cost} = \frac{\text{Discount percent}}{100 - \text{Discount percent}} \times \frac{360}{N}$$

where N = the number of days payment can be delayed by forgoing the cash discount. This equals the number of days credit is outstanding less the discount period.

The numerator of the first term (discount percent) is the cost per dollar of credit, whereas the denominator (100 – discount percent) represents the money available by forgoing the cash discount. The second term represents the number of times this cost is incurred in a year.

If you elect not to pay within the discount period, you should hold on to the money as long as possible. For example, if the terms are 2/10, net/60, you should not pay for 60 days.

Example 8.10 — The opportunity cost of *not* taking a discount when the terms are 3/15, net/60 is computed as follows:

$$\text{Opportunity cost} = \frac{3}{100-3} \times \frac{360}{60-15} = \frac{3}{97} \times \frac{360}{45} = 24.7\%$$

Table 8.1 presents possible credit terms and the associated opportunity cost of not paying within the discount period.

8.7 VOLUME DISCOUNTS

A *volume discount* is a reduction in the price you pay if a large quantity of goods is ordered. A larger order usually provides a higher trade discount.

Although a trade discount reduces the cost of buying goods, it can increase the carrying cost of holding a greater amount of merchandise.

Example 8.11 — If you buy 10,000 units instead of 5,000 units, you will receive a 2% discount off the purchase of $2 per unit. However, holding the higher level of merchandise will increase carrying costs by $300. The larger order is warranted as indicated below.

Savings due to trade discount:	
10,000 × 0.04, or ($2 × .02)	$400
Less: Increase in carrying costs	(300)
Net advantage of higher value order	$100

TABLE 8.1
Opportunity Cost of Not Taking a Discount

Credit Terms	Opportunity Cost
1/10, net/45	10.3%
2/10, net/30	36.7%
3/30, net/90	18.6%

8.8 CONCLUSION

By properly managing your cash, you can increase your rate of return. You should attempt to accelerate cash receipts and delay cash payments as much as is feasible.

9 How to Manage Your Accounts Receivable

Most business transactions are carried out on a credit basis. As a manager, your decisions regarding accounts receivable must include whether to give credit, and, if so, you must determine eligibility, amounts, and terms. The type of credit terms extended to customers determines the length of time a *customer* has to *pay* for the purchase. This, in turn, affects sales volume. For example, a longer credit term will probably result in increased sales, while a shorter credit term will likely result in less sales.

To be successful in managing accounts receivable, you must consider whether it is financially prudent to hold receivable balances. The opportunity cost of tying up money in accounts receivable is the loss of return that could have been earned if those funds had been invested in marketable securities. The management of these accounts receivable also involves having appropriate credit and collection policies.

The credit terms have a direct bearing on the associated costs and revenue to be generated from accounts receivable. For example, if credit terms are tight, there will be less investment in accounts receivable and less bad debt losses, but there will also be lower sales and reduced profits. On the other hand, if credit terms are lax, there will be higher sales and gross profit but greater bad debts and a higher opportunity cost of carrying the investment in accounts receivable because marginal customers take longer to pay.

In evaluating a potential customer's ability to pay, consider the customer's integrity, financial soundness, and collateral to be pledged. Bad debt losses can be estimated reliably when you sell to many customers and when credit policies have not changed for a long time.

You must also consider the costs of giving credit, including administrative costs of the credit department, computer services, fees to rating agencies, and periodic field investigations.

9.1 CREDIT REFERENCES

Retail credit bureaus and professional credit reference services should be used to appraise a customer's ability to pay. One such service is Dun and Bradstreet (D&B) reports, which contain information about a company's nature of business, product line, management, financial statement information, number of employees, previous payment history as reported by suppliers, amounts currently owed and past due, terms of sale, audit opinion, lawsuits, insurance coverage, leases, criminal proceedings, banking relationships and account information (e.g., current bank loans), location of business, and seasonality aspects. This information is provided in several

D&B publications such as the *D&B Reference Book*, the *D&B Consolidated Reports*, and *D&B Business Information Reports*.

9.2 CREDIT POLICY

A good credit system has the following characteristics:

1. It is straightforward, clear, consistent, and uniformly applied.
2. It is quick — otherwise, the customer may do business elsewhere.
3. It does not intrude into a customer's personal affairs.
4. It is inexpensive (e.g., there is centralization of credit decisions by experienced staff).
5. It is based upon past experience (e.g., examining the characteristics of good accounts, marginal accounts, delinquent accounts, and outright rejected applications). Careful attention is given to the reasons for previous uncollectibility. Table 9.1 shows an illustrative classification of customers by credit risk for a hypothetical company.

In managing accounts receivable, you should practice the following procedures. First, establish a *credit policy.*

1. A detailed review of a potential customer's soundness should be made prior to extending credit. The customer's financial statements and credit rating should be examined, and financial service reports should be reviewed. The customer's previous repayment record, the competitive factors of the customer's business, and the economy must also be considered. You want to know whether there is sufficient cash flow to repay the debt.
2. As customer financial health changes, credit limits should be revised.
3. Marketing factors must be noted since an excessively restricted credit policy will lead to lost sales.
4. If seasonal datings are used, you may offer liberal payments during slow periods to stimulate business by selling to customers who are unable to pay until later in the season. This policy is financially appropriate when the return on the additional sales plus the lowering in inventory costs is greater than the incremental cost of the additional investment in accounts receivable.
5. If the credit standing of a customer is dubious, collateral equal to or greater than the balance of the account should be pledged.
6. Avoid high-risk receivables (e.g., customers in a financially troubled industry or region).
7. Note that consumer receivables have a greater risk of default than corporate receivables. Individuals have less money than companies. Further, consumers are more protected by fair trade laws.

Competition and economic conditions must be considered. In recession, you may relax the credit policy because you need the business. However, in times of short supply, credit policy may be tightened because the seller is in the "driver's seat."

TABLE 9.1
Types of Customers and Expected Bad Debts

Risk Category	Description	Bad Debt Percent
A	Large, financially strong companies having minimal credit risk	None
B	Financially strong companies lacking a proven track record	2%
C	Companies with financial status of some risk and on which credit limit should be placed	3%
D	Marginal accounts needing reacted credit and careful monitoring	5%

Second, establish a *billing policy.*

1. Customer statements should be sent within 1 day subsequent to the close of the period.
2. Large sales should be billed immediately.
3. Customers should be invoiced for goods when the order is processed rather than when it is shipped.
4. Billing for services should be done on an interim basis or immediately prior to the actual services. The billing process will be more uniform if cycle billing is employed.

Finally, establish a *collection policy.*

1. Accounts receivable should be aged to identify delinquent and high-risk customers. *Aging* is determining the length of time an account is past due. The aging should be compared to industry norms. The longer receivables are outstanding, the higher the likelihood of uncollectibility.
2. Collection efforts should be undertaken at the very first sign of customer financial unsoundness. Use collection agencies when warranted.
3. Sell (factor) accounts receivable when net savings occur (see Chapter 16 for a more detailed discussion of factoring). Figure 9.1 shows the factoring procedure.
4. Credit insurance will protect against unusual bad debt losses.

9.3 ANALYZING ACCOUNTS RECEIVABLE

The average sales volume affects the size of the investment in accounts receivable. A larger quantity results in a larger investment. However, holding sales volume constant, the size of that investment becomes directly attributable to how long average receivables are outstanding. The calculation that defines this relationship follows:

Annual credit sales × Days to collect/360 = Investment in accounts receivable

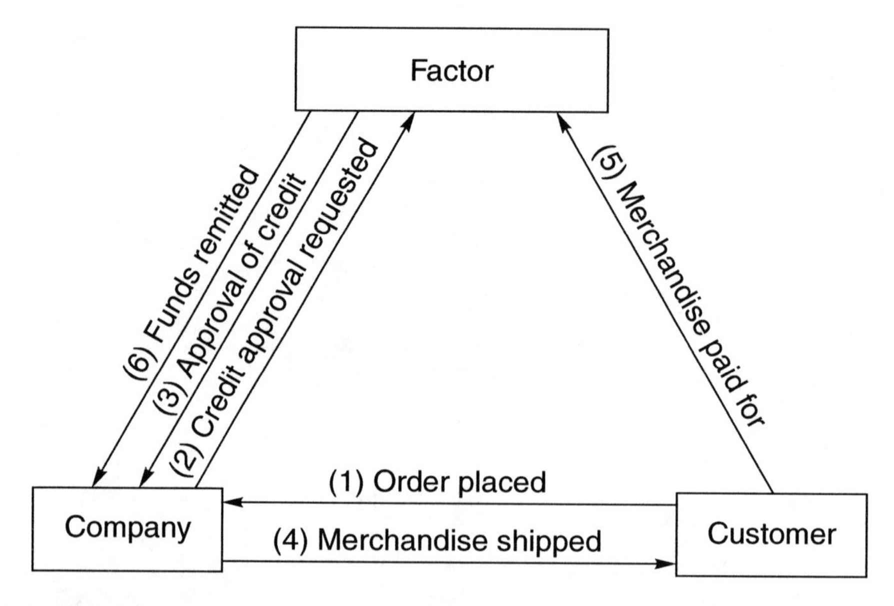

FIGURE 9.1 Factoring procedure.

Example 9.1 — You sell on terms of net/60. On the average, accounts are 30 days past due. Annual credit sales are $400,000. The investment in accounts receivable is

$$\$400{,}000 \times \frac{90}{360} = \$100{,}000$$

Example 9.2 — The cost of a product is 40% of the selling price, and the carrying cost is 12% of the selling price. On the average, accounts are paid 90 days after the sale date. Sales average $40,000 per month. Your accounts receivable for this product are calculated as follows:

$$3 \text{ months} \times \$40{,}000 \text{ sales} = \$120{,}000$$

The investment in accounts receivable is

$$\$120{,}000 \times (0.40 + 0.12) = \$62{,}400$$

Example 9.3 — You have accounts receivable of $700,000. The average manufacturing cost is 40% of the sales price. The before-tax profit margin is 10%. The carrying cost of inventory is 3% of the selling price. The sales commission is 8% of the sales. The investment in accounts receivable is

$$\$700{,}000 \times (0.\ 40 + 0.03 + 0.08) = \$700{,}000 \times (0.\ 51) = \$357{,}000$$

Example 9.4 — Your credit sales are $120,000, the collection period is 60 days, and the cost is 80% of sales price. The accounts receivable turnover is

$$\frac{360}{60} = 6$$

The average accounts receivable balance is

$$\frac{\text{Credit sales}}{\text{Turnover}} = \frac{\$120{,}000}{6} = \$20{,}000$$

The average investment in accounts receivable is

$$\$20{,}000 \times .80 = \$16{,}000$$

It is good business practice to give a discount for early payment by customers when the return on the funds received is greater than the cost of the discount.

Example 9.5 — Your current annual credit sales are $12,000,000. You sell on terms of net/30, and your collection period is 2 months. Therefore, the turnover of accounts receivable is 6 times (12/2). You could expect a 15% rate of return. You propose to offer a 3/10, net/30 discount. You anticipate 25% of your customers will take advantage of the discount. As a result of the discount policy, the collection period will be reduced to 11 months (turnover of 8 times). An analysis of this discount policy indicates that it would not be financially prudent.

Current average accounts receivable balance ($12,000,000/6)	$2,000,000
Average accounts receivable balance-after policy change ($12,000,000/8)	1,500,000
Reduction in average accounts receivable	500,000
Rate of return	× .15
Dollar return earned	75,000
Cost of discount (0.25 × $12,000,000 × 0.03)	90,000
Disadvantage of discount policy	$15,000

Instead, you may consider offering credit to customers with a higher-than-normal risk rating. Here, the profitability on additional sales must be compared with the amount of additional bad debts expected, higher investing and collection costs, and the opportunity cost of tying up funds in receivables for a longer period of time. Or you may consider other alternatives at slow business periods. When idle capacity exists, the additional profitability is represented by the incremental contribution margin (sales less variable costs) since fixed costs remain the same (see Chapter 3). The incremental investment in receivables is determined by multiplying the average accounts receivable by the ratio of per unit cost to selling price.

Example 9.6 — Currently, you sell 300,000 units annually, and each unit has a selling price of $80, a variable cost of $50, and a fixed cost of $10. The collection period is 2 months, and the rate of return is 16%. You are experiencing a period of idle capacity and are considering a change in policy that will relax credit standards. The following information applies to the proposal:

1. Sales will increase by 20%.
2. The collection period will increase to 3 months.
3. Bad debt losses are expected to be 3% of the increased sales.
4. Collection costs are expected to increase by $20,000.

An analysis of the proposed credit policy shows incremental profitability to be calculated as follows:

Increased unit sales (300,000 × .20)	60,000
Per-unit contribution margin ($80 – $50)	× $30
Incremental profit	$1,800,000

Additional bad debts are calculated as follows:

Incremental dollar sales (60,000 × $80)	$4,800,000
Bad debt percentage	× .03
Additional bad debts	$ 144,000

The new average unit cost is determined by using the following information:

	Units	Unit Cost	Total Cost
Current	300,000	$60	$18,000,000
Increment	60,000	50[a]	3,000,000
Total	360,000		$21,000,000

[a] Since idle capacity exists, the per-unit cost on the incremental sales is solely the variable cost of $50.

$$\text{New average unit cost} = \frac{\$21{,}000{,}000}{360{,}000} = \$58.33$$

The additional cost of higher investment in average accounts receivable is determined by calculating the investment in average accounts receivable after the change in policy:

$$\frac{\text{Credit sales}}{\text{Turnover}} \times \frac{\text{Unit cost}}{\text{Selling price}}$$

$$\frac{\$28{,}000{,}000}{4} \times \frac{\$58.33}{\$80.00} = \$5{,}249{,}700$$

From this amount, you must subtract the current investment in average accounts receivable:

$$\frac{\$24{,}000{,}000}{6} \times \frac{\$60}{\$80} = \$3{,}000{,}000$$

This results in the following information:

Incremental investment in average accounts receivable ($5,249,700 – $3,000,000)	$2,249,700
Rate of return	× .16
Additional cost	$ 359,952

The net advantage/disadvantage is

Incremental profitability		$1,800,000
Less: Additional bad debts	$144,000	
Additional collection costs	20,000	
Opportunity cost	359,952	
	523,952	523,952
Net advantage/disadvantage		$1,276,048

Since the net advantage is considerable, you should relax the credit policy.

Example 9.7 — You are considering liberalizing the credit policy to encourage more customers to purchase on credit. Currently, 80% of your sales are made on credit, and there is a gross margin of 30%. Other relevant data are included here.

	Current	**Proposed**
Sales	$300,000	$450,000
Credit sales	$240,000	$360,000
Collection expenses	4% of credit sales	5% of credit sales
Accounts receivable turnover	4.5	3

An analysis of the proposal yields the following results for average accounts receivable balance (credit sales/accounts receivable turnover):

Expected average accounts receivable ($360,000/3)	$120,000
Current average accounts receivable ($240,000/4.5)	53,333
Increase	$66,667

For gross profit, the analysis yields the following results:

Expected increase in credit sales ($360,000 – $240,000)	$120,000
Gross profit rate	× 0.30
Increase	$ 36,000

Collection expenses are calculated as follows:

Expected collection expenses (0.05 × $360, 000)	$18,000
Current collection expenses (0.04 × $240, 000)	9,600
Increase	$ 8,400

You would benefit from a more liberal credit policy.

9.4 CONCLUSION

You should carefully evaluate the credit and collection policies of your department to be certain that your accounts receivable are being aggressively managed. For example, during times of idle capacity, it may be advantageous to alter credit policies to attract business. Similarly, it might be necessary to tighten policies at other times in order to make better use of investment opportunities. Decisions regarding when to extend credit to whom and for how much must be made in light of these changing circumstances.

10 How to Manage Inventory

When managing inventory, remember that you want to avoid placing too many funds in inventory because it will result in declining profitability and retarded cash inflow. But some inventory must be maintained to meet customer orders and ensure smooth production activity. A sales forecast is the starting point for effective inventory management since expected sales determines how much inventory is needed.

The two types of inventory systems are *perpetual* and *periodic*. For the former, inventory balances are updated daily. For the latter, there is a physical count of inventory each period. Table 10.1 illustrates the perpetual inventory method, whereas Table 10.2 presents the periodic inventory method.

The three types of inventories are: (1) raw materials (materials acquired from a supplier that will be used in the manufacture of goods); (2) working-process (partially completed goods); and (3) finished goods (completed goods awaiting sale).

10.1 INVENTORY MANAGEMENT CONSIDERATIONS

In managing inventory, you should perform the following tasks:

1. Appraise the adequacy of the raw materials level, which depends on expected production, condition of equipment, and any seasonal considerations.
2. Forecast future movements in raw materials prices so that, if prices are expected to increase, additional materials are purchased at the lower price.
3. Discard slow-moving products to reduce inventory carrying costs and to improve cash flow.
4. Guard against inventory buildup, since it is associated with substantial carrying and opportunity costs.
5. Minimize inventory levels when liquidity and/or inventory financing problems exist.
6. Plan for a stock balance that will guard against and cushion the possible loss of business from a shortage in materials. The timing of an order also depends on seasonal factors.
7. Examine the quality of merchandise received. Monitoring the ratio of purchase returns to purchases should be enlightening: A sharp increase in the ratio indicates that a new supplier may be warranted.
8. Keep a careful record of back orders. A high back-order level indicates that fewer inventory balances are needed. The reason is that back orders may be used as indicators of the production required, resulting in improved production planning and procurement. The trend in the ratio of the dollar amount of back orders to the average per-day sales will prove useful.
9. Appraise the acquisition and inventory control functions. Any problems must be identified and rectified. In areas where control is weak, inventory balances should be restricted.

TABLE 10.1
Perpetual Inventory (Product X, units)

Date	Receipts	Issuances	Balance
8/1			10,000
8/5		8,000	2,000
8/10	10,000		12,000
8/13		6,000	6,000
8/17		4,000	2,000
8/29	10,000		12,000
8/30		4,000	8,000

TABLE 10.2
Periodic Inventory (Product X, Units)

August 1 Balance	10,000
Add: August purchases	20,000
Total stock available	30,000
Less: August 31 physical count	(8,000)
Sold or used during August	22,000

10. Closely supervise warehouse and materials handling staff to guard against theft and to maximize efficiency.
11. Minimize the lead time in the acquisition, manufacturing, and distribution functions. The lead time is how long it takes to receive merchandise from suppliers after an order is placed. Depending upon lead times, an increase in inventory stocking may be required or the purchasing pattern may have to be altered.
12. Examine the time between raw materials input and the completion of production to see if production and engineering techniques can be implemented to hasten the production operation.
13. Examine the degree of spoilage and take steps to reduce it.
14. Maintain proper inventory control, such as through the application of computer techniques. For example, a point-of-sale computerized electronic register may be used by a retail business. The register continually updates inventory for sales and purchases. These data facilitate the computation of reorder points and quantity per order.
15. Scrutinize the trend in the unit cost of manufactured items. Reasons for variations should be analyzed to see if they are caused by factors within or beyond your control (i.e., due to an increase in oil prices or to managerial inefficiencies).
16. Have economies in production run size to *reduce setup costs* and *idle* time.

10.2 INVENTORY ANALYSIS

You must consider the obsolescence and spoilage risk of inventory. For example, technological, perishable, fashionable, flammable, and specialized goods usually have a high salability risk. The nature of the risk should be taken into account in computing the desired inventory level for the item.

Different inventory items vary in profitability and the amount of space they take up. Inventory management involves a trade-off between the costs of keeping inventory vs. the benefits of holding it. Higher inventory levels result in increased costs from storage, casualty and theft insurance, spoilage, higher property taxes for larger facilities, increased manpower requirements, and interest on borrowed funds to finance inventory acquisition. However, an increase in inventory lowers the possibility of lost sales from stockouts and the incidence of production slowdowns from inadequate inventory. Additionally, large volume purchases will result in greater purchase discounts. Inventory levels are also affected by short-term interest rates. For instance, as short-term interest rates increase, the optimum level of holding inventory will be reduced.

You may have to decide whether it is more profitable to sell inventory as is or to sell it after further processing. Example 10.1 provides a hypothetical situation that involves making this decision.

Example 10.1 — You can sell inventory for $40,000 as is or sell it for $80,000 if you put $20,000 into further processing. The latter should be selected because further processing yields a $60,000 profit relative to $40,000 for the current sale.

Inventory should be counted at regular, cyclic intervals because this method allows you to check inventory on an ongoing basis as well as to reconcile the book and physical amounts. Cyclic counting has the following advantages:

1. Permits the efficient use of a few, full-time experienced counters throughout the year.
2. Enables the timely detection and correction of the causes of inventory error.
3. Does not require a plant shutdown, as does a year-end count.
4. Facilitates the modification of computer inventory programs, if needed.

A quantity discount may be received when purchasing large orders. The discount reduces the acquisition cost of materials, as Example 10.2 shows.

Example 10.2 — You purchase 2,000 units of an item with a list price of $8 each. The quantity discount is 4%. The net cost of the item is calculated as follows:

Acquisition cost (2,000 × $8)	$16,000
Less: Discount (.04 × $16,000)	(640)
Net cost	$15,360

The average investment in inventory should be considered, as Example 10.3 illustrates.

Example 10.3 — You place an order for 7,000 units at the beginning of the year. Each unit costs $12. The average investment is

Average inventory[a]	3,500
Unit cost	× $12
Average investment	$42,000

[a] $\frac{\text{Quantity (Q)}}{2} = \frac{7{,}000}{2} = 3{,}500$

The more frequently you place an order, the lower the average investment will be.

An additional item should be held as long as the increase in sales revenue from storing the item equals the cost of holding it. In inventory planning, you also have to consider your sales volume and cost estimates.

10.3 DETERMINING THE CARRYING AND ORDERING COSTS

Inventory carrying costs include warehousing, handling, insurance, and property taxes. A provisional cost for spoilage and obsolescence should also be included in an analysis of inventory. Further, the opportunity cost of holding inventory balances should be taken into account. The more the inventory held, the greater its carrying cost. Carrying cost is calculated as follows:

$$\text{Carrying cost} = \frac{Q}{2} \times C$$

where $Q/2$ represents average quantity and C is the carrying cost per unit.

Inventory ordering costs are the costs of placing an order and receiving the merchandise. They include freight and the clerical costs. If you want to minimize ordering costs, enter the fewest number of orders possible. In the case of produced items, ordering cost includes scheduling cost. Ordering cost is formulated as follows:

$$\text{Ordering cost} = \frac{S}{Q} \times P$$

where S = total usage
Q = quantity per order
P = cost of placing an order

Example 10.4 uses this equation in a hypothetical situation.

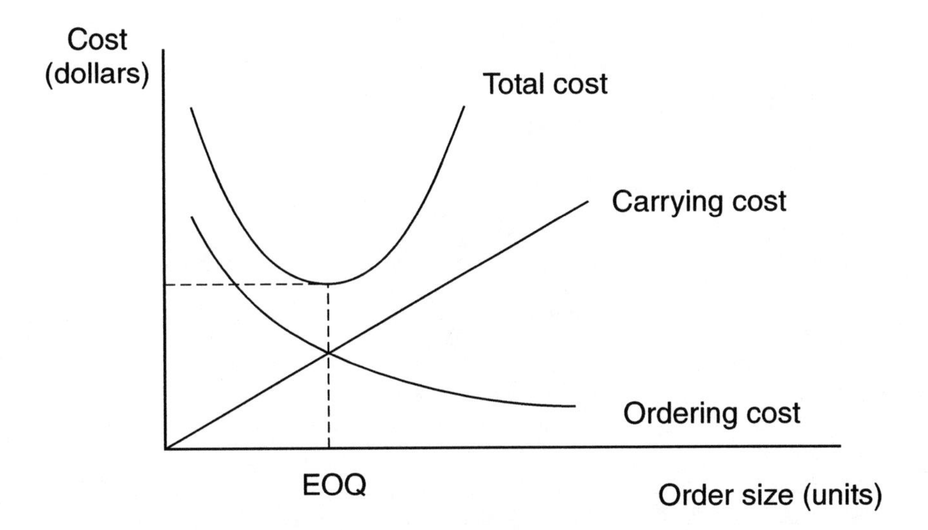

FIGURE 10.1 EOQ point.

Example 10.4 — You use 8,500 units per year. Each order is for 200 units. The cost per order is \$13. The total ordering cost for the year is:

$$\frac{8{,}500}{200} \times \$13 = \$552.50$$

The total inventory cost is therefore

$$\frac{QC}{2} + \frac{SP}{C}$$

A trade-off exists between ordering and carrying costs. A greater order quantity will increase carrying costs but lower ordering costs.

10.4 THE ECONOMIC ORDER QUANTITY (EOQ)

The economic order quantity (EOQ) is the optimum amount of goods to order each time so that total inventory costs are minimized. You should apply EOQ analysis to every product that represents a significant proportion of sales. EOQ is calculated as follows:

$$\text{EOQ} = \sqrt{\frac{2SP}{C}}$$

The number of orders for a period is the usage, S, divided by the EOQ.

Figure 10.1 graphs the EOQ point, while Examples 10.5 and 10.6 calculate the frequency of ordering for hypothetical situations.

Example 10.5 — You want to know how frequently to place orders. The following information is provided:

S = 500 units per month
P = \$40 per order
C = \$4 per unit

$$\text{EOQ} = \sqrt{\frac{2SP}{C}} = \sqrt{\frac{2(500)(40)}{4}} = \sqrt{10{,}000} = 100 \text{ units}$$

The number of orders each month is

$$\text{S/EOQ} = 500/100 = 5$$

Therefore, an order should be placed about every 6 days (31/5).

Example 10.6 — A store is determining its frequency of orders for blenders. Each blender costs \$15. The annual carrying cost is \$200. The ordering cost is \$10. The store anticipates selling 50 blenders each month. Its desired average inventory level is 40.

$S = 50 \times 12 = 600$
$P = \$10$

$$C = \frac{\text{Purchase price} \times \text{Carrying cost}}{\text{Average investment}} = \frac{\$15 \times \$200}{40 \times \$15} = \$5$$

$$\text{EOQ} = \sqrt{\frac{2SP}{C}} = \sqrt{\frac{2(600)(10)}{5}} = \sqrt{\frac{12{,}000}{5}} = \sqrt{2{,}400} = 49 \text{ (rounded)}$$

The number of orders per year is:

$$\frac{S}{\text{EOQ}} = \frac{600}{49} = 12 \text{ orders (rounded)}$$

The store should place an order about every 30 days (365/12).

During periods of inflation and tight credit, you should be flexible with your inventory management policies. For example, the EOQ computation will have to be adjusted for increasing costs.

10.5 AVOIDING STOCKOUTS

Stockout of raw materials or work-in-process can cause a slowdown in production. To avoid a stockout situation, a safety stock should be kept. Safety stock is the minimum inventory for an item, based on expected usage and delivery time of materials. This cushions against unusual product demand or unexpected delivery problems. Safety stock helps to prevent the potential damage to customer relations and to future sales that can occur when you lack the inventory to fill an order immediately. The need for a safety stock increases the total inventory required.

Example 10.7 illustrates the safety stock principle.

Example 10.7 — You place an order when the inventory level reaches 210 units rather than 180 units. The safety stock is 30 units. In other words, you expect to be stocked with 30 units when the new order is received.

The optimum safety stock is the point where the increased carrying cost equals the opportunity cost of a potential stockout. The increased carrying cost equals the carrying cost per unit multiplied by the safety stock:

$$\text{Stockout cost} = \text{Number of orders} \times \left(\frac{\text{usage}}{\text{order quantity}}\right) \times \text{Stockout units}$$

$$\times \text{ Unit stockout cost} \times \text{Probability of a stockout}$$

Example 10.8 uses the safety stock equation in a hypothetical situation.

Example 10.8 — You use 100,000 units annually. Each order calls for 10,000 units. Stockout is 1,000 units; this amount is the difference between the maximum daily usage during the lead time less the reorder point, ignoring a safety stock factor. You expect a stockout probability of 30%. The per-unit stockout cost is \$2.30. The carrying cost per unit is \$5. The stockout cost and the amount of safety stock are determined as follows:

$$\text{Stockout cost} = \frac{100{,}000}{10{,}000} \times 1{,}000 \times \$2.30 \times .3 = \$6{,}900$$

Let X represent the safety stock.

$$\text{Stockout cost} = \text{Carrying cost of safety stock}$$

$$\$6{,}900 = \$5X$$

$$1{,}380 \text{ units} = X$$

10.6 DETERMINING THE REORDER POINT OR ECONOMIC ORDER POINT (EOP)

The reorder point tells you when to place an order. However, calculating the reorder point requires you to know the lead time from placing to receiving an order. Thus, the reorder point is computed as follows:

$$\text{EOP} = \text{Lead time} \times \text{Average usage per unit of time}$$

This tells you the inventory level at which a new order should be placed. If you need a safety stock, then add this amount to the EOP.

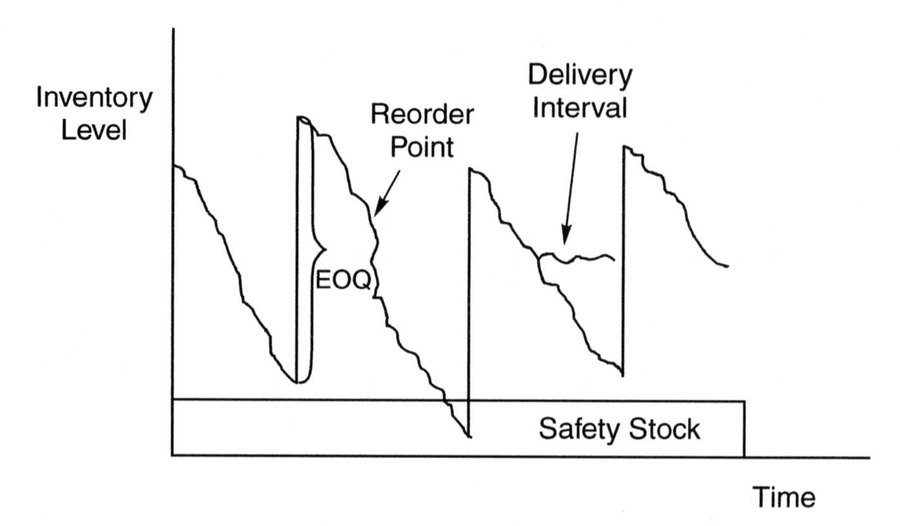

FIGURE 10.2 Changes in level of inventory over time.

Example 10.9 — You need 6,400 units evenly throughout the year. There is a lead time of 1 week. There are 50 working weeks in the year. The reorder point is

$$1 \text{ week} \times \frac{6{,}400}{50 \text{ weeks}} = 1 \text{ week} \times 128 \text{ units/week} = 128 \text{ units}$$

Figure 10.2 shows the changes in level of inventory over time. An inventory control model appears in Figure 10.3.

We now turn to determining what inventory items should have priority in space. The product that results in the highest contribution margin per cubic foot should typically have the most priority in space allocation.

Example 10.10 — For Product X:

Sales price	$ 15
Variable cost per unit	5
Contribution margin per unit	$ 10
Contribution margin per dozen	$120
Storage space per dozen (cubic feet)	÷ 2
Contribution margin per cubic foot	$ 60
Number of units sold per period	× 4
Expected contribution margin per cubic foot	$240

10.7 THE ABC INVENTORY CONTROL METHOD

The ABC method of inventory control requires the classification of inventory into one of three groups — A, B, or C. Group A items are the most expensive, group B less expensive, and group C the least costly. The higher the value of the inventory items, the more control needed. See Figure 10.4 for an illustration of the ABC inventory control system. Table 10.3 illustrates an ABC distribution.

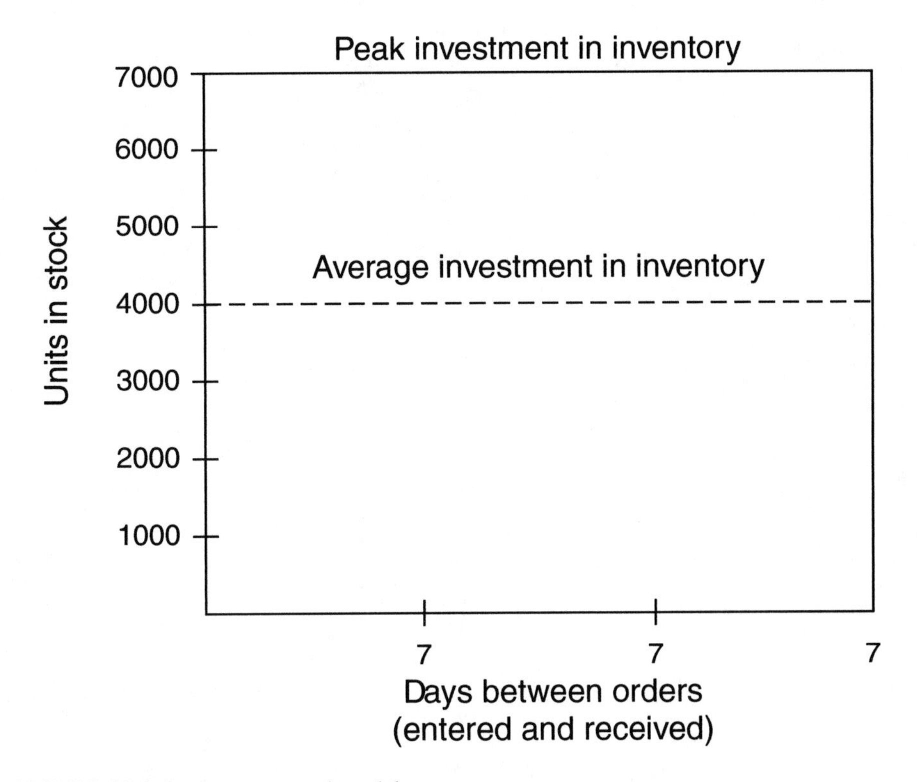

FIGURE 10.3 Inventory control model.

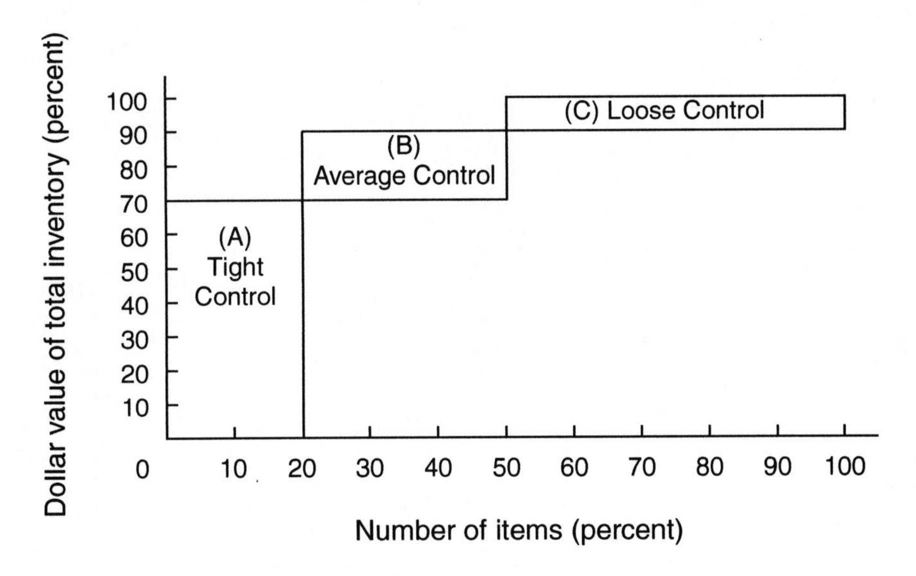

FIGURE 10.4 ABC inventory control system.

TABLE 10.3
ABC Inventory Distribution

Inventory Classification	Population (Percent)	Dollar Usage (Percent)
A	20	80
B	30	15
C	50	5

10.8 CONCLUSION

Deciding on the proper amount of investment in inventory is not an easy task. The amount you need may change daily and require close evaluation. However, improper inventory management can unnecessarily tie up money in inventory funds that can be used more productively elsewhere. A buildup of inventory may lead to obsolete inventory. On the other hand, an excessively low inventory may result in less profit because of lost sales.

Part III

Financial Decision Making for Managers

11 Understanding the Concept of Time Value

A dollar now is worth more than a dollar to be received later. This statement sums up an important principle: money has a time value. The truth of this principle is not that inflation might make the dollar received at a later time worth less in buying power. The reason is that you could invest the dollar now and have more than a dollar at the specified later date.

Time value of money is a critical consideration in financial and investment decisions. For example, compound interest calculations are needed to determine future sums of money resulting from an investment. Discounting, or the calculation of present value, which is inversely related to compounding, is used to evaluate the future cash flow associated with capital budgeting projects. There are plenty of applications of time value of money in finance. For example, how do you determine the periodic payout of an auto loan or a mortgage loan? This chapter discusses the concepts, calculations, and applications of future values and present values.

11.1 FUTURE VALUES — HOW MONEY GROWS

A dollar in hand today is worth more than a dollar to be received tomorrow because of the interest it could earn from putting it in a savings account or placing it in an investment account. Compounding interest means that interest earns interest. For the discussion of the concepts of compounding and time value, let us define the following:

F_n = future value = the amount of money at the end of year n
P = principal
i = annual interest rate
n = number of years

Then,

F_1 = the amount of money at the end of year 1
= principal and interest = $P + iP = P(1 + i)$
F_2 = the amount of money at the end of year 2
= $F_1(1 + i) = P(1 + i)(1 + i) = P(1 + i)^2$

The future value of an investment compounded annually at rate i for n years is

$$F_n = P(1 + i)^n = P \times T_1(i,n)$$

where $T_1(i,n)$ is the compounded amount of \$1 and can be found in Table 11.1 at the end of the chapter.

Example 11.1 — You place $1,000 in a savings account earning 8% interest compounded annually. How much money will you have in the account at the end of 4 years?

$$F_n = P(1 + i)^n$$

$$F_4 = \$1{,}000\ (1 + 0.08)^4 = \$1{,}000\ T_1(8\%, 4\ \text{years})$$

From Table 11.1, the T_1 for 4 years at 8% is 1.361. Therefore,

$$F_4 = \$1{,}000\ (1.361) = \$1{,}361.$$

Example 11.2 — You invested a large sum of money in the stock of TLC Corporation. The company paid a $3 dividend per share. The dividend is expected to increase by 20% per year for the next 3 years. You wish to project the dividends for years 1 through 3.

$$F_n = P(1 + i)^n$$

$$F_1 = \$3(1 + 0.2)^1 = \$3\ T_1(20\%,1) = \$3\ (1.200) = \$3.60$$

$$F_2 = \$3(1 + 0.2)^2 = \$3\ T_1(20\%,2) = \$3\ (1.440) = \$4.32$$

$$F_3 = \$3(1 + 0.2)^3 = \$3\ T_1(20\%,3) = \$3\ (1.728) = \$5.18$$

11.2 INTRAYEAR COMPOUNDING

Interest is often compounded more frequently than once a year. Banks, for example, compound interest quarterly, daily, and even continuously. If interest is compounded m times a year, then the general formula for solving the future value becomes

$$F_n = P\left(1 + \frac{i}{m}\right)^{n \times m} = P \times T_1(i/m,\ n \cdot m)$$

The formula reflects more frequent compounding ($n \cdot m$) at a smaller interest rate per period (i/m). For example, in the case of semiannual compounding ($m = 2$), the above formula becomes

$$F_n = P(1 + i/2)^{n \times 2} = P \times T_1(i/2,\ n \cdot 2)$$

Example 11.3 — You deposit $10,000 in an account offering an annual interest rate of 20%. You will keep the money on deposit for 5 years. The interest rate is compounded quarterly. The accumulated amount at the end of the fifth year is calculated as follows:

$$F_n = P\left(1 + \frac{i}{m}\right)^{n \times m} = P \times T_1(i/m,\ n \cdot m)$$

where

$$P = \$10,000$$
$$i/m = 20\%/4 = 5\%$$
$$n \times m = 5 \times 4 = 20$$

Therefore,

$$F_5 = \$10,000(1 + 0.05)^{20} = \$10,000\ T_1(5\%,20) = \$10,000\ (2.653) = \$26,530$$

Example 11.4 — Assume that P = \$1,000, i = 8% and n = 2 years. Then for annual compounding (m = 1):

$$F_2 = \$1,000(1 + 0.08)^2 = \$1,000\ T_1(8\%,2) = \$1,000(1.166) = \$1,166$$

Semiannual compounding (m = 2):

$$F_2 = \$1,000\left(1 + \frac{0.08}{2}\right)^{2\times 2}$$
$$= \$1,000(1 + .04)^4 = \$1,000\ T_1(4\%,4)$$
$$= \$1,000(1.170)$$
$$= \$1,170$$

Quarterly compounding (m = 4):

$$F_2 = \$1,000\left(1 + \frac{0.08}{4}\right)^{2\times 4}$$
$$= \$1,000(1 + .02)^8 = \$1,000\ T_1(2\%,8)$$
$$= \$1,000(1.172)$$
$$= \$1,172$$

As the example shows, the more frequently interest is compounded, the greater the amount accumulated. This is true for any interest for any period of time.

11.3 FUTURE VALUE OF AN ANNUITY

An annuity is defined as a series of payments (or receipts) of a fixed amount for a specified number of periods. Each payment is assumed to occur at the end of the period. The future value of an annuity is a compound annuity, which involves depositing or investing an equal sum of money at the end of each year for a certain number of years and allowing it to grow.

Let S_n = the future value on an n-year annuity

A = the amount of an annuity

Then we can write

$$S_n = A(1 + i)^{n-1} + A(1 + i)^{n-2} + \ldots + A(1 + i)^0$$
$$= A[(1 + i)^{n-1} + (1 + i)^{n-2} + \ldots + (1 + i)^0]$$
$$= A \times \sum_{t=0}^{n-1} (1+i)^t = A\left[\frac{(1+i)^n - 1}{i}\right] = A \times T_2(i, n)$$

where $T_2(i,n)$ represents the future value of an annuity of $1 for n years compounded at i percent and can be found in Table 11.2 at the end of the chapter.

Example 11.5 — You wish to determine the sum of money you will have in a savings account at the end of 6 years by depositing $1,000 at the end of each year for the next 6 years. The annual interest rate is 8 percent. The T_2(8%,6 years) is given in Table 11.2 as 7.336. Therefore,

$$S_6 = \$1{,}000\ T_2(8\%,6) = \$1{,}000\ (7.336) = \$7{,}336$$

Example 11.6 — You deposit $30,000 semiannually into a fund for 10 years. The annual interest rate is 8%. The amount accumulated at the end of the tenth year is calculated as follows:

$$S_n = A \times T_2(i,n)$$

where

$$A = \$30{,}000$$
$$i = 8\%/2 = 4\%$$
$$n = 10 \times 2 = 20$$

Therefore,

$$S_n = \$30{,}000\ T_2(4\%,20)$$
$$= \$30{,}000\ (29.778) = \$893{,}340$$

11.4 PRESENT VALUE — HOW MUCH IS MONEY WORTH NOW?

Present value is the present worth of future sums of money. The process of calculating present values, or discounting, is actually the opposite of finding the compounded future value. In connection with present value calculations, the interest rate i is called the *discount rate*.

Recall that $F_n = P\ (1 + i)^n$. Therefore,

$$P = \frac{F_n}{(1+i)^n} = F_n\left[\frac{1}{(1+i)^n}\right] = F_n \times T_3(i,n)$$

where $T_3(i,n)$ represents the present value of \$1 and is given in Table 11.3 at the end of the chapter.

Example 11.7 — You have been given an opportunity to receive \$20,000 6 years from now. If you can earn 10% on your investments, what is the most you should pay for this opportunity? To answer this question, you must compute the present value of \$20,000 to be received 6 years from now at a 10% rate of discount. F_6 is \$20,000, i is 10%, and n is 6 years. $T_3(10\%,6)$ from Table 11.3 is 0.565.

$$P = \$2,000\left[\frac{1}{(1+0.1)^6}\right] = \$20,000\ T_3(10\%,6) = \$20,000(0.565) = \$11,300$$

This means that you can earn 10% on your investment, and you would be indifferent to receiving \$11,300 now or \$20,000 6 years from today since the amounts are time equivalent. In other words, you could invest \$11,300 today at 10 percent and have \$20,000 in 6 years.

11.5 PRESENT VALUE OF MIXED STREAMS OF CASH FLOWS

The present value of a series of mixed payments (or receipts) is the sum of the present value of each individual payment. We know that the present value of each individual payment is the payment times the appropriate T_3 value.

Example 11.8 — You are thinking of starting a new product line that initially costs \$32,000. Your annual projected cash inflows are:

Year 1 \$10,000
Year 2 \$20,000
Year 3 \$5,000

If you must earn a minimum of 10% on your investment, should you undertake this new product line?

The present value of this series of mixed streams of cash inflows is calculated as follows:

Year	Cash inflows	×	$T_3(10\%, n)$	Present value
1	\$10,000		0.909	\$9,090
2	\$20,000		0.826	16,520
3	\$5,000		0.751	3,755
				\$29,365

Since the present value of your projected cash inflows is less than the initial investment, you should not undertake this project.

11.6 PRESENT VALUE OF AN ANNUITY

Interest received from bonds, pension funds, and insurance obligations all involve annuities. To compare these financial instruments, we need to know the present value of each. The present value of an annuity (P_n) can be found by using the following equation:

$$P_n = A \times \left[\frac{1}{(1+i)^1}\right] + A \times \left[\frac{1}{(1+i)^2}\right] + \ldots + A \times \left[\frac{1}{(1+i)^n}\right]$$

$$= A \times \left[\frac{1}{(1+i)^1} + \frac{1}{(1+i)^2} + \ldots + \frac{1}{(1+i)^n}\right]$$

$$= A \times \sum_{t=1}^{n} \frac{1}{(1+i)^t} \sum A \times \frac{1}{i}\left[1 - \frac{1}{(1+i)}\right]$$

$$= A \times T_4(i,n)$$

where $T_4(i,n)$ represents the present value of an annuity of $1 discounted at i percent for n years and is found in Table 11.4 at the end of the chapter.

Example 11.9 — Assume that the cash inflows in Example 11.8 form an annuity of $10,000 for 3 years. Then the present value is

$$P_n = A \times T_4(i,n)$$

$$P_3 = \$10{,}000\ T_4(10\%,\ 3\ \text{years}) = \$10{,}000\ (2.487) = \$24{,}870$$

Example 11.10 — Suppose that you have just won the state lottery in the amount of $1 million (or $800,000 after taxes). Instead of paying you the lump sum of $800,000, the state pays you $40,000 each year for the next 20 years. If the discount rate is 10%, how much is the lottery amount worth to you?

Then the present worth is

$$P_n = A \times T_4(i,n)$$

$$P_{20} = \$40{,}000 \times T_4\ (10\%,\ 20\ \text{years}) = \$40{,}000\ (8.514) = \$340{,}560$$

This means that if the state can make a 10% return on its lottery sales receipts, it is actually paying you only $340,560 rather than $800,000.

11.7 PERPETUITIES

Some annuities go on forever, called *perpetuities*. An example of a perpetuity is preferred stock that yields a constant dollar dividend indefinitely. The present value of a perpetuity is found as follows:

$$\text{Present value of a perpetuity} = \frac{\text{receipt}}{\text{discount rate}} = \frac{A}{i}$$

Example 11.11 — Assume that a perpetual bond has an $80-per-year interest payment and that the discount rate is 10%. The present value of this perpetuity is:

$$P = \frac{A}{i} = \frac{\$80}{0.10} = \$800$$

11.8 APPLICATIONS OF FUTURE VALUES AND PRESENT VALUES

Future and present values have numerous applications in financial and investment decisions. Six of these applications are presented below.

11.9 DEPOSITS TO ACCUMULATE A FUTURE SUM (OR SINKING FUND)

An individual might wish to find the annual deposit (or payment) that is necessary to accumulate a future sum. To find this future amount (or sinking fund) we can use the formula for finding the future value of an annuity.

$$S_n = A \times T_2(i,n)$$

Solving for A, we obtain:

$$\text{Annual deposit amount} = A = \frac{S_n}{T_2(i,n)}$$

Example 11.12 — You wish to determine the equal annual end-of-year deposits required to accumulate $5,000 at the end of 5 years in a fund. The interest rate is 10%. The annual deposit is:

$$S_5 = \$5{,}000$$

$$T_2(10\%, 5 \text{ years}) = 6.105 \text{ (from Table 11.2)}$$

$$A = \frac{\$5{,}000}{6.105} = \$819$$

In other words, if you deposit $819 at the end of each year for 5 years at 10% interest, you will have accumulated $5,000 at the end of the fifth year.

Example 11.13 — You need a sinking fund for the retirement of a bond 30 years from now. The interest rate is 10%. The annual year-end contribution needed to accumulate $1,000,000 is:

$$S_{30} = \$1,000,000$$

$$T_2(10\%, 30 \text{ years}) = 164.49$$

$$A = \frac{\$1,000,000}{164.49} = 6,079.40$$

11.10 AMORTIZED LOANS

If a loan is to be repaid in equal periodic amounts, it is said to be an amortized loan. Examples include auto loans, mortgage loans, and most commercial loans. The periodic payment can easily be computed as follows:

$$P_n = A \times T_4(i,n)$$

Solving for A, we obtain:

$$\text{Amount of loan} = A = \frac{P_n}{T_4(i,n)}$$

Example 11.14 — You borrow $200,000 for 5 years at an interest rate of 14%. The annual year-end payment on the loan is calculated as follows:

$$P_5 = \$200,000$$

$$T_4(14\%, 5 \text{ years}) = 3.433 \text{ (from Table 11.4)}$$

$$\text{Amount of loan} = A = \frac{P_5}{T_4(14\%, 5 \text{ years})} = \frac{\$200,000}{3.433} = \$58,258.08$$

Example 11.15 — You take out a 40-month bank loan of $5,000 at a 12% annual interest rate. You want to find out the monthly loan payment.

$$i = 12\%/12 \text{ months} = 1\%$$

$$P_{40} = \$5,000$$

$$T_4(1\%, 40 \text{ months}) = 32.835 \text{ (from Table 11.4)}$$

$$\text{Therefore, } A = \frac{\$5,000}{32.835} = \$152.28$$

So, to repay the principal and interest on a $5,000, 12%, 40-month loan, you have to pay $152.28 a month for the next 40 months.

Example 11.16 — Assume that a firm borrows $2,000 to be repaid in three equal installments at the end of each of the next 3 years. The bank charges 12% interest. The amount of each payment is

$$P_3 = \$2,000$$

$$T_4(12\%, 3 \text{ years}) = 2.402$$

Therefore, $A = \dfrac{\$2,000}{2.402} = \$832.64$

Each loan payment consists partly of interest and partly of principal. The breakdown is often displayed in a loan amortization schedule. The interest component of the payment is largest in the first period (because the principal balance is the highest) and subsequently declines, whereas the principal portion is smallest in the first period (because of the high interest) and increases thereafter, as shown in the following example.

Example 11.17 — Using the same data as in Example 15, we set up the following amortization schedule:

Year	Payment	Interest	Repayment of Principal	Remaining Balance
0				$2,000.00
1	$832.64	$240.00[a]	$592.64[b]	$1,407.36
2	$832.64	$168.88	$663.76	$ 743.60
3	$832.64	$89.23	$743.41[c]	

[a] Interest is computed by multiplying the loan balance at the beginning of the year by the interest rate. Therefore, interest in year 1 is $2,000(0.12) = $240; in year 2 interest is $1,407.36(0.12) = $168.88; and in year 3 interest is $743.60(0.12) = $89.23. All figures are rounded.
[b] The reduction in principal equals the payment less the interest portion ($832.64 – $240.00 = $592.64).
[c] Not exact because of accumulated rounding errors.

11.11 ANNUAL PERCENTAGE RATE (APR)

Different types of investments use different compounding periods. For example, most bonds pay interest semiannually; banks generally pay interest quarterly. If a financial manager wishes to compare investments with different compounding periods, he or she needs to put them on a common basis. The annual percentage rate (APR), or effective annual rate, is used for this purpose and is computed as follows:

$$APR = \left(1 + \frac{i}{m}\right)^m - 1.0$$

where i = the stated, nominal, or quoted rate and m = the number of compounding periods per year.

Example 11.18 — If the nominal rate is 6%, compounded quarterly, the APR is

$$APR = \left(1 + \frac{i}{m}\right)^m - 1.0 = \left(1 + \frac{0.06}{4}\right)^4 - 1.0 = (1.015)^4 - 1.0 = 1.0614 - 1.0 = 0.0614 = 6.14\%$$

This means that if one bank offered 6% with quarterly compounding, while another offered 6.14% with annual compounding, they would both be paying the same effective rate of interest.

Annual percentage rate (APR) also is a measure of the cost of credit, expressed as a yearly rate. It includes interest as well as other financial charges such as loan origination and certain closing fees. The lender is required to tell you the APR. It provides you with a good basis for comparing the cost of loans, including mortgage plans.

11.12 RATES OF GROWTH

In finance, it is necessary to calculate the compound annual rate of growth, associated with a stream of earnings. The compound annual growth rate in earnings per share is computed as follows:

$$F_n = P \times T_1(i,n)$$

Solving this for T_1, we obtain

$$T_1(i,n) = \frac{F_n}{P}$$

Example 11.19 — Assume that your company has earnings per share of $2.50 in 2001, and 10 years later the earnings per share has increased to $3.70. The compound annual rate of growth in earnings per share is computed as follows:

$$F_{10} = \$3.70 \text{ and } P = \$2.50$$

Therefore,

$$T_1(i,10) = \frac{\$3.70}{\$2.50} = 1.48$$

From Table 11.1 a T_1 of 1.48 at 10 years is at i = 4%. The compound annual rate of growth is therefore 4%.

11.13 COMPOUND ANNUAL RATE OF INTEREST

The compound annual interest rate is computed as follows:

$$F_n = P \times T_1(i,n)$$

$$\text{or } S_n = P \times T_2(i,n)$$

Solving this for T_1 (or T_2), we obtain

$$T_1(i,n) = \frac{F_n}{P}$$

$$\text{or } T_2(i,n) = \frac{S_n}{P}$$

Example 11.20 — You agree to pay back $3,000 in 6 years on a $2,000 loan made today. You are being charged an interest rate of 7%. Thus,

$$T_1(i,n) = \frac{F_n}{P}$$

$$T_1(i, 6 \text{ years}) = \frac{\$3{,}000}{\$2{,}000} = 1.5$$

so that i = 7% (from Table 11.1).

Example 11.21 — You want to have $500,000 accumulated in a pension plan after 9 years. You deposit $30,000 per year. Thus,

$$T_2(i,n) = \frac{S_n}{P}$$

$$T_2(i, 9 \text{ years}) = \frac{\$500{,}000}{\$30{,}000} = 16.667$$

so that i = 15% (approximately, from Table 11.2).

11.14 BOND VALUES

Bonds call for the payment of a specific amount of interest for a stated number of years and the repayment of the face value at the maturity date. Thus, a bond represents an annuity plus a lump sum. Its value is found as the present value of the payment stream. The interest is usually paid semiannually.

$$V = \sum_{t=1}^{n} \frac{I}{(1+i)^t} + \frac{M}{(1+i)^n}$$

$$= I \times T_4(i,n) + M \times T_3(i,n)$$

where

I = interest payment per period
M = par value, or maturity value, usually $1,000
i = investor's required rate of return
n = number of periods

Example 11.22 — Assume there is a 10-year bond with a 10% coupon, paying interest semiannually and having a face value of $1,000. Since interest is paid semiannually, the number of periods involved is 20 and the semiannual cash inflow is $100/2 = $50.

Assume that you have a required rate of return of 12% for this type of bond. Then, the present value (V) of this bond is:

$$V = \$50 \cdot T_4(6\%, 20) + \$1{,}000 \times T_3(6\%, 20)$$
$$= \$50(11.470) + \$1{,}000(0.312) = \$573.50 + \$312.00 = \$885.50$$

Note that the required rate of return (12%) is higher than the coupon rate of interest (10%), so the bond value (or the price investors are willing to pay for this particular bond) is less than its $1,000 face value.

11.15 USE OF FINANCIAL CALCULATORS AND SPREADSHEET PROGRAMS

There are many financial calculators that contain preprogrammed formulas to perform many present value and future applications. They include Radio Shack EC5500, Hewlett-Packard 10B, Sharp EL733, and Texas Instruments BA35. Furthermore, spreadsheet software such as Microsoft Excel has built-in financial functions to perform many such applications. For example, PMT (*principal, interest, term*) in Lotus 1-2-3 or Excel calculates the amount of the periodic payment to payoff a loan, given a specified periodic interest rate and number of payment periods.

11.16 CONCLUSION

The basic idea of the time value of money is that money received in the future is not as valuable as money received today. The time value of money is a critical factor in many financial and investment applications, such as finding the amount of deposits to accumulate a future sum and the periodic payment of an amortized loan. The development of the time value of money concept permits comparison of sums of money that are available at different points in time. This chapter developed two basic concepts: future value and present value. It showed how these values are calculated and can be applied to various financial and investment situations.

TABLE 11.1
The Future Value of $1.00 (Compounded Amount of $1.00)
$(1 + i)^n = T_1(i, n)$

Periods	4%	6%	8%	10%	12%	14%	20%
1	1.040	1.060	1.080	1.100	1.120	1.140	1.200
2	1.082	1.124	1.166	1.210	1.254	1.300	1.440
3	1.125	1.191	1.260	1.331	1.405	1.482	1.728
4	1.170	1.263	1.361	1.464	1.574	1.689	2.074
5	1.217	1.338	1.469	1.611	1.762	1.925	2.488
6	1.265	1.419	1.587	1.772	1.974	2.195	2.986
7	1.316	1.504	1.714	1.949	2.211	2.502	3.583
8	1.369	1.594	1.851	2.144	2.476	2.853	4.300
9	1.423	1.690	1.999	2.359	2.773	3.252	5.160
10	1.480	1.791	2.159	2.594	3.106	3.707	6.192
11	1.540	1.898	2.332	2.853	3.479	4.226	7.430
12	1.601	2.012	2.518	3.139	3.896	4.818	8.916
13	1.665	2.133	2.720	3.452	4.364	5.492	10.699
14	1.732	2.261	2.937	3.798	4.887	6.261	12.839
15	1.801	2.397	3.172	4.177	5.474	7.138	15.407
16	1.873	2.540	3.426	4.595	6.130	8.137	18.488
17	1.948	2.693	3.700	5.055	6.866	9.277	22.186
18	2.026	2.854	3.996	5.560	7.690	10.575	26.623
19	2.107	3.026	4.316	6.116	8.613	12.056	31.948
20	2.191	3.207	4.661	6.728	9.646	13.743	38.338
30	3.243	5.744	10.063	17.450	29.960	50.950	237.380
40	4.801	10.286	21.725	45.260	93.051	188.880	1469.800

TABLE 11.2
The Future Value of an Annuity of $1.00[a] (Compounded Amount of an Annuity of $1.00) $\frac{(1+i)^n - 1}{i} = T_2(i, n)$

Periods	4%	6%	8%	10%	12%	14%	20%
1	1.000	1.000	1.000	1.000	1.000	1.000	1.000
2	2.040	2.060	2.080	2.100	2.120	2.140	2.200
3	3.122	3.184	3.246	3.310	3.374	3.440	3.640
4	4.247	4.375	4.506	4.641	4.779	4.921	5.368
5	5.416	5.637	5.867	6.105	6.353	6.610	7.442
6	6.633	6.975	7.336	7.716	8.115	8.536	9.930
7	7.898	8.394	8.923	9.487	10.089	10.730	12.916
8	9.214	9.898	10.637	11.436	12.300	13.233	16.499
9	10.583	11.491	12.488	13.580	14.776	16.085	20.799
10	12.006	13.181	14.487	15.938	17.549	19.337	25.959
11	13.486	14.972	16.646	18.531	20.655	23.045	32.150
12	15.026	16.870	18.977	21.385	24.133	27.271	39.580
13	16.627	18.882	21.495	24.523	28.029	32.089	48.497
14	18.292	21.015	24.215	27.976	32.393	37.581	59.196
15	20.024	23.276	27.152	31.773	37.280	43.842	72.035
16	21.825	25.673	30.324	35.950	42.753	50.980	87.442
17	23.698	28.213	33.750	40.546	48.884	59.118	105.930
18	25.645	30.906	37.450	45.600	55.750	68.394	128.120
19	27.671	33.760	41.446	51.160	63.440	78.969	154.740
20	29.778	36.778	45.762	57.276	75.052	91.025	186.690
30	56.085	79.058	113.283	164.496	241.330	356.790	1181.900
40	95.026	154.762	259.057	442.597	767.090	1342.000	7343.900

[a] Payments (or receipts) at the *end* of each project.

TABLE 11.3
Present Value of $1.00 $\frac{1}{(1+i)^n} = T_3(i, n)$

Periods	4%	6%	8%	10%	12%	14%	16%	18%	20%	22%	24%	26%	28%	30%	40%
1	.962	.943	.926	.909	.893	.877	.862	.847	.833	.820	.806	.794	.781	.769	.714
2	.925	.890	.857	.826	.797	.769	.743	.718	.694	.672	.650	.630	.610	.592	.510
3	.889	.840	.794	.751	.712	.675	.641	.609	.579	.551	.524	.500	.477	.455	.364
4	.855	.792	.735	.683	.636	.592	.552	.516	.482	.451	.423	.397	.373	.350	.260
5	.822	.747	.681	.621	.567	.519	.476	.437	.402	.370	.341	.315	.291	.269	.186
6	.790	.705	.630	.564	.507	.456	.410	.370	.335	.303	.275	.250	.227	.207	.133
7	.760	.665	.583	.513	.452	.400	.354	.314	.279	.249	.222	.198	.178	.159	.095
8	.731	.627	.540	.467	.404	.351	.305	.266	.233	.204	.179	.157	.139	.123	.068
9	.703	.592	.500	.424	.361	.308	.263	.225	.194	.167	.144	.125	.108	.094	.048
10	.676	.558	.463	.386	.322	.270	.227	.191	.162	.137	.116	.099	.085	.073	.035
11	.650	.527	.429	.350	.287	.237	.195	.162	.135	.112	.094	.079	.066	.056	.025
12	.625	.497	.397	.319	.257	.208	.168	.137	.112	.092	.076	.062	.052	.043	.018
13	.601	.469	.368	.290	.229	.182	.145	.116	.093	.075	.061	.050	.040	.033	.013
14	.577	.442	.340	.263	.205	.160	.125	.099	.078	.062	.049	.039	.032	.025	.009
15	.555	.417	.315	.239	.183	.140	.108	.084	.065	.051	.040	.031	.025	.020	.006
16	.534	.394	.292	.218	.163	.123	.093	.071	.054	.042	.032	.025	.019	.015	.005
17	.513	.371	.270	.198	.146	.108	.080	.060	.045	.034	.026	.020	.015	.012	.003
18	.494	.350	.250	.180	.130	.095	.069	.051	.038	.028	.021	.016	.012	.009	.002
19	.475	.331	.232	.164	.116	.083	.060	.043	.031	.023	.017	.012	.009	.007	.002
20	.456	.312	.215	.149	.104	.073	.051	.037	.026	.019	.014	.010	.007	.005	.001
21	.439	.294	.199	.135	.093	.064	.044	.031	.022	.015	.011	.008	.006	.004	.001
22	.422	.278	.184	.123	.083	.056	.038	.026	.018	.013	.009	.006	.004	.003	.001
23	.406	.262	.170	.112	.074	.049	.033	.022	.015	.010	.007	.005	.003	.002	
24	.390	.247	.158	.102	.066	.043	.028	.019	.013	.008	.006	.004	.003	.002	
25	.375	.233	.146	.092	.059	.038	.024	.016	.010	.007	.005	.003	.002	.001	
26	.361	.220	.135	.084	.053	.033	.021	.014	.009	.006	.004	.002	.002	.001	
27	.347	.207	.125	.076	.047	.029	.018	.011	.007	.005	.003	.002	.001	.001	
28	.333	.196	.116	.069	.042	.026	.016	.010	.006	.004	.002	.002	.001	.001	
29	.321	.185	.107	.063	.037	.022	.014	.008	.005	.003	.002	.001	.001	.001	
30	.308	.174	.099	.057	.033	.020	.012	.007	.004	.003	.002	.001	.001		
40	.208	.097	.046	.022	.011	.005	.003	.001	.001						

TABLE 11.4
Present Value of an Annuity of $1.00[a] $\frac{1}{i}\left[1-\frac{1}{(1+i)^n}\right] = T_4(i,n)$

Periods	4%	6%	8%	10%	12%	14%	16%	18%	20%	22%	24%	25%	26%	28%	30%	40%
1	0.962	0.943	0.926	0.909	0.893	0.877	0.862	0.847	0.833	0.820	0.806	0.800	0.794	0.781	0.769	0.714
2	1.886	1.833	1.783	1.736	1.690	1.647	1.605	1.566	1.528	1.492	1.457	1.440	1.424	1.392	1.361	1.224
3	2.775	2.673	2.577	2.487	2.402	2.322	2.246	2.174	2.106	2.042	1.981	1.952	1.923	1.868	1.816	1.589
4	3.630	3.465	3.312	3.170	3.037	2.914	2.798	2.690	2.589	2.494	2.404	2.362	2.320	2.241	2.166	1.849
5	4.452	4.212	3.993	3.791	3.605	3.433	3.274	3.127	2.991	2.864	2.745	2.689	2.635	2.532	2.436	2.035
6	5.242	4.917	4.623	4.355	4.111	3.889	3.685	3.498	3.326	3.167	3.020	2.951	2.885	2.759	2.643	2.168
7	6.002	5.582	5.206	4.868	4.564	4.288	4.039	3.812	3.605	3.416	3.242	3.161	3.083	2.937	2.802	2.263
8	6.733	6.210	5.747	5.335	4.968	4.639	4.344	4.078	3.837	3.619	3.421	3.329	3.241	3.076	2.925	2.331
9	7.435	6.802	6.247	5.759	5.328	4.946	4.607	4.303	4.031	3.786	3.566	3.463	3.366	3.184	3.019	2.379
10	8.111	7.360	6.710	6.145	5.650	5.216	4.833	4.494	4.192	3.923	3.682	3.571	3.465	3.269	3.092	2.414
11	8.760	7.887	7.139	6.495	5.938	5.453	5.029	4.656	4.327	4.035	3.776	3.656	3.544	3.335	3.147	2.438
12	9.385	8.384	7.536	6.814	6.194	5.660	5.197	4.793	4.439	4.127	3.851	3.725	3.606	3.387	3.190	2.456
13	9.986	8.853	7.904	7.103	6.424	5.842	5.342	4.910	4.533	4.203	3.912	3.780	3.656	3.427	3.223	2.468
14	10.563	9.295	8.244	7.367	6.628	6.002	5.468	5.008	4.611	4.265	3.962	3.824	3.695	3.459	3.249	2.477
15	11.118	9.712	8.559	7.606	6.811	6.142	5.575	5.092	4.675	4.315	4.001	3.859	3.726	3.483	3.268	2.484
16	11.652	10.106	8.851	7.824	6.974	6.265	5.669	5.162	4.730	4.357	4.033	3.887	3.751	3.503	3.283	2.489
17	12.166	10.477	9.122	8.022	7.120	6.373	5.749	5.222	4.775	4.391	4.059	3.910	3.771	3.518	2.395	2.492
18	12.659	10.828	9.372	8.201	7.250	6.467	5.818	5.273	4.812	4.419	4.080	3.928	3.786	3.529	3.304	2.494
19	13.134	11.158	9.604	8.365	7.366	6.550	5.877	5.316	4.844	4.442	4.097	3.942	3.799	3.539	3.311	2.496
20	13.590	11.470	9.818	8.514	7.469	6.623	5.929	5.353	4.870	4.460	4.110	3.954	3.808	3.546	3.316	2.497

TABLE 11.4 (continued)

Periods	4%	6%	8%	10%	12%	14%	16%	18%	20%	22%	24%	25%	26%	28%	30%	40%
21	14.029	11.764	10.017	8.649	7.562	6.687	5.973	5.384	4.891	4.476	4.121	3.963	3.816	3.551	3.320	2.498
22	14.451	12.042	10.201	8.772	7.645	6.743	6.011	5.410	4.909	4.488	4.130	3.970	8.822	3.556	3.323	2.498
23	14.857	12.303	10.371	8.883	7.718	6.792	6.044	5.432	4.923	4.499	4.137	3.976	3.827	3.559	3.325	2.499
24	15.247	12.550	10.529	8.985	7.784	6.835	6.073	5.451	4.937	4.507	4.143	3.981	3.831	3.562	3.327	2.499
25	15.622	12.783	10.675	9.077	7.843	6.873	6.097	5.467	4.948	4.514	4.147	3.985	3.834	3.564	3.329	2.499
26	15.983	13.003	10.810	9.161	7.896	6.906	6.118	5.480	4.956	4.520	4.151	3.988	3.837	3.566	3.300	2.500
27	16.330	13.211	10.935	9.237	7.943	6.935	6.136	5.492	4.964	4.524	4.154	3.990	3.839	3.567	3.331	2.500
28	16.663	13.406	11.051	9.307	7.984	6.961	6.152	5.502	4.970	4.528	4.157	3.992	3.840	3.568	3.331	2.500
29	16.984	13.591	11.158	9.370	8.022	6.983	6.166	5.510	4.975	4.531	4.159	3.994	3.841	3.569	3.332	2.500
30	17.292	13.765	11.258	9.427	8.055	7.003	6.177	5.517	4.979	4.534	4.160	3.995	3.842	3.569	3.332	2.500
40	19.793	15.046	11.925	9.779	8.244	7.105	6.234	5.548	4.997	4.544	4.166	3.999	3.846	3.571	3.333	2.500

[a] Payments (or receipts) at the *end* of each period.

12 Capital Investment Decisions

Capital budgeting is the process of making long-term investment decisions. These decisions should be made in light of the company's goals. The stockholders have entrusted the company with their money and they expect the firm to invest it wisely. Investments in fixed assets should be consistent with the goal of maximizing the firm's market value.

There are many investment decisions that the company may have to make in order to grow. Examples of capital budgeting applications are product line selection, keep-or-sell a business segment decisions, lease or buy decisions, and determination of which assets to invest in. To make long-term investment decisions in accordance with your goal, you must perform three tasks in evaluating capital budgeting projects: (1) estimate cash flows; (2) estimate the cost of capital (or required rate of return); and (3) apply a decision rule to determine if a project is "good" or "bad."

This chapter discusses:

- The types and special features of capital budgeting decisions.
- Basic capital budgeting techniques.
- How to select the best mix of projects with a limited capital spending budget.
- How income tax factors affect investment decisions.
- The types of depreciation methods.
- The effect of the Modified Accelerated Cost Recovery System (MACRS) on capital budgeting decisions.
- How to compute a firm's cost of capital.

12.1 WHAT ARE THE TYPES OF INVESTMENT PROJECTS?

There are typically two types of long-term investment decisions made by your company:

1. *Selection decisions* in terms of obtaining new facilities or expanding existing facilities. Examples include:
   (a) Investments in property, plant, and equipment as well as other types of assets.
   (b) Resource commitments in the form of new product development, market research, introduction of a computer, refunding of long-term debt, etc.
   (c) Mergers and acquisitions in the form of buying another company to add a new product line. Also see Chapter 17.
2. *Replacement decisions* in terms of replacing existing facilities with new facilities. Examples include replacing an old machine with a high-tech machine.

12.2 WHAT ARE THE FEATURES OF INVESTMENT PROJECTS?

Long-term investments have three important features:

1. They typically involve a large amount of initial cash outlays that tend to have a long-term impact on the firm's future profitability. Therefore, these initial cash outlays need to be justified on a cost-benefit basis.
2. There are expected recurring cash inflows (for example, increased revenues, savings in cash operating expenses, etc.) over the life of the investment project. This frequently requires considering the *time value of money.*
3. Income taxes could make a difference in the accept or reject decision. Therefore, income tax factors must be taken into account in every capital budgeting decision.

12.3 HOW DO YOU MEASURE INVESTMENT WORTH?

Several methods of evaluating investment projects are as follows:

1. Payback period
2. Accounting rate of return (ARR)
3. Net present value (NPV)
4. Internal rate of return (IRR)
5. Profitability index (or cost/benefit ratio)

The NPV method and the IRR method are called *discounted cash flow* (DCF) methods. Each of these methods is discussed below.

12.3.1 PAYBACK PERIOD

The payback period measures the length of time required to recover the amount of initial investment. It is computed by dividing the initial investment by the cash inflows through increased revenues or cost savings.

Examples 12.1 and 12.2 calculate the payback periods for two different situations.

Example 12.1 — Consider the following data:

Cost of investment	$18,000
Annual after-tax cash savings	$3,000

Then, the payback period is formulated as follows:

$$\text{Payback period} = \frac{\text{Initial investment}}{\text{Cost savings}} = \frac{\$18{,}000}{\$3{,}000} = 6 \text{ years}$$

Decision rule: Choose the project with the shorter payback period. The rationale behind this choice is: The shorter the payback period, the less risky the project, and the greater the liquidity.

Example 12.2 — Consider the two projects whose after-tax cash inflows are not even. Assume each project costs $1,000.

	Cash Inflow	
Year	**A($)**	**B($)**
1	100	500
2	200	400
3	300	300
4	400	100
5	500	
6	600	

When cash inflows are not even, the payback period has to be found by trial and error. The payback period of project A is ($1,000= $100 + $200 + $300 + $400) 4 years. The payback period of project B is ($1,000 = $500 + $400 + $100):

$$2 \text{ years} + \frac{\$100}{\$300} = 2\tfrac{1}{3} \text{ years}$$

Project B is the project of choice in this case, since it has the shorter payback period.

The advantages of using the payback period method of evaluating an investment project are that (1) it is simple to compute and easy to understand, and (2) it handles investment risk effectively. The shortcomings of this method are that (1) it does not recognize the time value of money, and (2) it ignores the impact of cash inflows received after the payback period; essentially, cash flows after the payback period determine profitability of an investment.

12.3.2 Accounting Rate of Return (ARR)

Accounting rate of return (ARR) measures profitability from the conventional accounting standpoint by relating the required investment — or sometimes the average investment — to the future annual net income.

Decision rule: Under the ARR method, choose the project with the higher rate of return.

Example 12.3 — Consider the following investment:

Initial investment	$6,500
Estimated life	20 years
Cash inflows per year	$1,000
Depreciation per year (using straight line method)	$325

The accounting rate of return for this project is:

$$\text{APR} = \frac{\text{Net income}}{\text{Investment}} = \frac{\$1{,}000 - \$325}{\$6{,}500} = 10.4\%$$

If average investment (usually assumed to be one-half of the original investment) is used, then:

$$\text{APR} = \frac{\$1{,}000 - \$325}{\$3{,}250} = 20.8\%$$

The advantages of this method are that it is easily understandable, is simple to compute, and recognizes the profitability factor.

The shortcomings of this method are that it fails to recognize the time value of money, and it uses accounting data instead of cash flow data.

12.3.3 Net Present Value (NPV)

Net present value (NPV) is the excess of the present value (PV) of cash inflows generated by the project over the amount of the initial investment (I):

$$\text{NPV} = \text{PV} - \text{I}$$

The present value of future cash flows is computed using the so-called cost of capital (or minimum required rate of return) as the discount rate. In the case of an annuity, the present value would be

$$\text{PV} = \text{A} \times \text{T}_4(\text{i, n})$$

where A is the amount of the annuity. The value of T_4 is found in Table 11.4.

Decision rule: If NPV is positive, accept the project. Otherwise, reject it.

Example 12.4 — Consider the following investment:

Initial investment	$12,950
Estimated life	10 years
Annual cash inflows	$3,000
Cost of capital (minimum required rate of return)	12%

Present value of the cash inflows is:

$\text{PV} = \text{A} \times \text{T}_4$ (i, n)	
$= \$3{,}000 \times \text{T}_4$ (12%, 10 years)	
= $3,000 (5.650)	$16,950
Initial investment (I)	12,950
Net present value (NPV = PV – I)	$ 4,000

Since the NPV of the investment is positive, the investment should be accepted.

The advantages of the NPV method are that it obviously recognizes the time value of money and it is easy to compute whether the cash flows form an annuity or vary from period to period.

12.3.4 Internal Rate of Return (IRR)

Internal rate of return (IRR) is defined as the rate of interest that equates I with the PV of future cash inflows. In other words, for an IRR,

$$I = PV$$

or

$$NPV = 0$$

Decision rule: Accept the project if the IRR exceeds the cost of capital. Otherwise, reject it.

Example 12.5 — Assume the same data given in Example 12.4, and set the following equality (I = PV):

$$\$12{,}950 = \$3{,}000 \times T_4(i, 10 \text{ years})$$

$$T_4(i, 10 \text{ years}) = \frac{\$12{,}950}{\$3{,}000} = 4.317$$

which stands somewhere between 18 and 20% in the 10-year line of Table 11.4. The interpolation follows:

	PV of an Annuity of $1 Factor $T_4(i,10 \text{ years})$	
18%	4.494	4.495
IRR	4.317	
20%		4.192
Difference	0.177	0.302

Therefore,

$$\text{IRR} = 18\% + \frac{0.177}{0.302}(20\% - 18\%)$$

$$= 18\% + 0.586(2\%) = 18\% + 1.17\% = 19.17\%$$

Since the IRR of the investment is greater than the cost of capital (12%), accept the project.

The advantage of using the IRR method is that it does consider the time value of money and, therefore, is more exact and realistic than the ARR method. The shortcomings of this method are that (1) it is time-consuming to compute, especially when the cash inflows are not even, although most business calculators have a program to calculate IRR, and (2) it fails to recognize the varying sizes of investment in competing projects.

When cash inflows are not even, IRR is computed by the trial-and-error method, which is not discussed here. Financial calculators such as Texas Instruments and Sharp have a key for IRR calculations.

12.3.5 Profitability Index

The profitability index is the ratio of the total PV of future cash inflows to the initial investment, that is, PV/I. This index is used as a means of ranking projects in descending order of attractiveness.

Decision rule: If the profitability index is greater than 1, then accept the project.

Example 12.6 — Using the data in Example 12.4, the profitability index is

$$\frac{PV}{I} = \frac{\$16{,}950}{\$12{,}950} = 1.31$$

Since this project generates $1.31 for each dollar invested (i.e., its profitability index is greater than 1), accept the project.

The profitability index has the advantage of putting all projects on the same relative basis regardless of size.

12.4 HOW TO SELECT THE BEST MIX OF PROJECTS WITH A LIMITED BUDGET

Many firms specify a limit on the overall budget for capital spending. Capital rationing is concerned with the problem of selecting the mix of acceptable projects that provides the highest overall NPV. The profitability index is used widely in ranking projects competing for limited funds.

Example 12.7 — A company with a fixed budget of $250,000 needs to select a mix of acceptable projects from the following:

Projects	I($)	PV($)	NPV($)	Profitability Index	Ranking
A	70,000	112,000	42,000	1.6	1
B	100,000	145,000	45,000	1.45	2
C	110,000	126,500	16,500	1.15	5
D	60,000	79,000	19,000	1.32	3
E	40,000	38,000	–2,000	0.95	6
F	80,000	95,000	15,000	1.19	4

The ranking resulting from the profitability index shows that the company should select projects A, B, and D.

	I	PV
A	$70,000	$112,000
B	100,000	145,000
D	60,000	79,000
	$230,000	$336,000

Therefore,

NPV = $336,000 – $230,000 = $106,000

12.5 HOW DO INCOME TAXES AFFECT INVESTMENT DECISIONS?

Income taxes make a difference in many capital budgeting decisions. In other words, the project that is attractive on a before-tax basis may have to be rejected on an after-tax basis. Income taxes typically affect both the amount and the timing of cash flows. Since net income, not cash inflows, is subject to tax, after-tax cash inflows are not usually the same as after-tax net income.

Let us define:

S = Sales
E = Cash operating expenses
d = Depreciation
t = Tax rate

Then,

Before-tax cash inflows (or before-tax *cash savings*) = S – E
and net income = S – E – d.

By definition,

After-tax cash inflows = Before-tax cash inflows – Taxes
= (S – E) – (S – E – d) (t)

Rearranging gives the short-cut formula:

After-tax cash inflows = (S – E) (1 – t) + (d)(t)

As can be seen, the deductibility of depreciation from sales in arriving at net income subject to taxes reduces income tax payments and thus serves as a tax shield.

Tax shield = Tax savings on depreciation = (d)(t)

Example 12.8 — Assume:

S = $12,000

E = $10,000

d = $500 per year using the straight line method

t = 30%

Then

After-tax cash inflow = ($12,000 – $10,000)(1 – 0.3) + ($500)(0.3)

= ($2,000)(0.7) + ($500)(0.3)

= $1,400 + $150 = $1,550

Note that a tax shield = tax savings on depreciation = (d)(t)

= ($500)(0.3) = $150

Since the tax shield is dt, the higher the depreciation deduction, the higher the tax savings on depreciation will be. Therefore, an accelerated depreciation method (such as double-declining balance) produces higher tax savings than the straight-line method. Accelerated methods produce higher present values for the tax savings that may make a given investment more attractive.

Example 12.9 — The Shalimar Company estimates that it can save $2,500 a year in cash operating costs for the next 10 years if it buys a special-purpose machine at a cost of $10,000. No salvage value is expected. Assume that the income tax rate is 30%, and the after-tax cost of capital (minimum required rate of return) is 10%. After-tax cash savings can be calculated as follows:

Note that depreciation by straight-line is $10,000/10 = $1,000 per year. Here before-tax cash savings = (S – E) = $2,500. Thus,

After-tax cash savings = (S – E) (1 – t) + (d)(t)

= $2,500(1 – 0.3) + $1,000(0.3)

= $1,750 + $300 = $2,050

To see if this machine should be purchased, the net present value can be calculated.

PV = $2,050 T_4(10%, 10 years) = $2,050 (6.145) = $12,597.25.

Thus, NPV = PV – I = $12,597.25 – $10,000 = $2,597.25
Since NPV is positive, the machine should be bought.

12.6 TYPES OF DEPRECIATION METHODS

We saw that depreciation provided the tax shield in the form of (d)(t). Among the commonly used depreciation methods are straight-line and accelerated methods. The two major accelerated methods are sum-of-the-years'-digits (SYD) and double-declining-balance (DDB).

12.6.1 STRAIGHT-LINE METHOD

This is the easiest and most popular method of calculating depreciation. It results in equal periodic depreciation charges. The method is most appropriate when an asset's usage is uniform from period to period, as is the case with furniture. The annual depreciation expense is calculated by using the following formula:

$$\text{Depreciation expense} = \frac{\text{Cost} - \text{Salvage value}}{\text{Number of years of useful life}}$$

Example 12.10 — An auto is purchased for \$20,000 and has an expected salvage value of \$2,000. The auto's estimated life is 8 years. Its annual depreciation is calculated as follows:

$$\text{Depreciation expense} = \frac{\text{Cost} - \text{Salvage value}}{\text{Number of years of useful life}}$$

$$= \frac{\$20{,}000 - \$2{,}000}{8 \text{ years}} = \$2{,}250 \text{ / year}$$

An alternative means of computation is to multiply the *depreciable* cost (\$18,000) by the annual depreciation rate, which is 12.5% in this example. The annual rate is calculated by dividing the number of years of useful life into one (1/8 = 12.5%). The result is the same: \$18,000 × 12.5% = \$2,250.

12.6.2 SUM-OF-THE-YEARS'-DIGITS (SYD) METHOD

In this method, the number of years of life expectancy is enumerated in reverse order in the numerator, and the denominator is the sum of the digits. For example, if the life expectancy of a machine is 8 years, write the numbers in reverse order: 8, 7, 6, 5, 4, 3, 2, 1. The sum of these digits is 36, or (8 + 7 + 6 + 5 + 4 + 3 + 2 + 1). Thus, the fraction for the first year is 8/36, while the fraction for the last year is 1/36. The sum of the eight fractions equals 36/36, or 1. Therefore, at the end of 8 years, the machine is completely written down to its salvage value.

The following formula may be used to quickly find the sum-of-the-years'-digits (S):

$$S = \frac{(N)(N+1)}{2}$$

where N represents the number of years of expected life.

Example 12.11 — In Example 12.10, the *depreciable* cost is $18,000 ($20,000 – $2,000). Using the SYD method, the computation for each year's depreciation expense is

$$S = \frac{(N)(N+1)}{2} = \frac{8(9)}{2} = \frac{72}{2} = 36$$

Year	Fraction	×	Depreciation Amount ($)	=	Depreciation Expense
1	8/36		$18,000		$ 4,000
2	7/36		18,000		3,500
3	6/36		18,000		3,000
4	5/36		18,000		2,500
5	4/36		18,000		2,000
6	3/36		18,000		1,500
7	2/36		18,000		1,000
8	1/36		18,000		500
Total					$18,000

12.6.3 Double-Declining-Balance (DDB) Method

Under this method, depreciation expense is highest in the earlier years and lower in the later years. First, a depreciation rate is determined by doubling the straight-line rate. For example, if an asset has a life of 10 years, the straight-line rate is 1/10 or 10%, and the double-declining rate is 20%. Second, depreciation expense is computed by multiplying the rate by the book value of the asset at the beginning of each year. Since book value declines over time, the depreciation expense decreases each successive period.

This method *ignores* salvage value in the computation. However, the book value of the fixed asset at the end of its useful life cannot be below its salvage value.

Example 12.12 — Assume the data in Example 12.10. Since the straight-line rate is 12.5% (1/8), the double-declining-balance rate is 25% (2 × 12.5%). The depreciation expense is computed as follows:

Year	Book Value at Beginning of Year	×	Rate (%)	=	Depreciation Expense	Year-end Book Value
1	$20,000		25%		$5,000	$15,000
2	15,000		25		3,750	11,250
3	11,250		25		2,813	8,437
4	8,437		25		2,109	6,328
5	6,328		25		1,582	4,746
6	4,746		25		1,187	3,559
7	3,559		25		667	2,002

Note: If the original estimated salvage value had been $2,100, the depreciation expense for the eighth year would have been $569 ($2,669 - $2,100) rather than $667, since the asset cannot be depreciated below its salvage value.

TABLE 12.1
Modified Accelerated Cost Recovery System Classification of Assets

	Property Class					
Year	3-year	5-year	7-year	10-year	15-year	20-year
1	33.3%	20.0%	14.3%	10.0%	5.0%	3.8%
2	44.5	32.0	24.5	18.0	9.5	7.2
3	14.8[a]	19.2	17.5	14.4	8.6	6.7
4	7.4	11.5[a]	12.5	11.5	7.7	6.2
5		11.5	8.9[a]	9.2	6.9	5.7
6		5.8	8.9	7.4	6.2	5.3
7			8.9	6.6[a]	5.9[a]	4.9
8			4.5	6.6	5.9	4.5[a]
9				6.5	5.9	4.5
10				6.5	5.9	4.5
11				3.3	5.9	4.5
12					5.9	4.5
13					5.9	4.5
14					5.9	4.5
15					5.9	4.5
16					3.0	4.4
17						4.4
18						4.4
19						4.4
20						4.4
21						2.2
Total	100%	100%	100%	100%	100%	100%

[a] Denotes the year of changeover to straight-line depreciation.

12.7 HOW DOES MACRS AFFECT INVESTMENT DECISIONS?

Although the traditional depreciation methods still can be used for computing depreciation for book purposes, 1981 saw a new way of computing depreciation deductions for tax purposes. That rule is called the *Modified Accelerated Cost Recovery System* (MACRS) rule, as enacted by Congress in 1981 and then modified somewhat under the Tax Reform Act of 1986. This rule is characterized as follows:

1. It abandons the concept of useful life and accelerates depreciation deductions by placing all depreciable assets into one of eight age property classes. It calculates deductions, based on an allowable percentage of the asset's original cost (see Tables 12.1 and 12.2). With a shorter life than useful life, the company would be able to deduct depreciation more quickly and save more in income taxes in the earlier years, thereby making an investment more attractive. The rationale behind the system is that this way the

TABLE 12.2
MACRS Tables by Property Class

MACRS Property Class and Depreciation Method	Useful Life (ADR Midpoint Life)[a]	Examples of Assets
3-year property 200% declining balance	4 years or less	Most small tools are included; the law specifically excludes autos and light trucks from this property class.
5-year property 200% declining balance	More than 4 years to less than 10 years	Autos and light trucks, computers, typewriters, copiers, duplicating equipment, heavy general-purpose trucks, and research and experimentation equipment are included.
7-year property 200% declining balance	10 years or more to less than 16 years	Office furniture and fixtures, most items of machinery and equipment used in production are included.
10-year property 200% declining balance	16 years or more to less than 20 years	Various machinery and equipment, such as that used in petroleum distilling and refining and in the milling of grain, are included.
15-year property 150% declining balance	20 years or more to less than 25 years	Sewage treatment plants, telephone and electrical distribution facilities, and land improvements are included.
20-year property 150% declining balance	25 years or more	Service stations and other real property with an ADR midpoint life of less than 27.5 years are included.
27.5-year property straight-line	Not applicable	All residential rental property is included.
31.5-year property straight-line	Not applicable	All nonresidential real property is included.

[a] The term ADR midpoint life means the "useful life" of an asset in a business sense; the appropriate ADR midpoint lives for assets are designated in the tax regulations.

government encourages the company to invest in facilities and increase its productive capacity and efficiency. (Remember that the higher d, the larger the tax shield (d)(t)).

2. Since the allowable percentages in Table 12.2 add up to 100%, there is no need to consider the salvage value of an asset in computing depreciation.
3. The company may elect the straight-line method. The straight-line convention must follow what is called the *half-year convention*. This means that the company can deduct only half of the regular straight-line depreciation amount in the first year. The reason for electing to use the MACRS optional straight-line method is that some firms may prefer to stretch out depreciation deductions using the straight-line method rather than accelerate them. Those firms are the ones that just start out, or have little or no income and wish to show more income on their income statements.

Example 12.13 — Assume that a machine falls under a 3-year property class and costs $3,000 initially. The straight-line option under MACRS differs from the traditional straight-line method in that under this method the company would deduct only $500 depreciation in the first year and the fourth year ($3,000/3 years = $1,000; $1,000/2 = $500). The table below compares the straight-line with half-year convention with the MACRS.

Year	Straight-Line (half-year) Depreciation	Cost		MACRS %	MACRS Deduction
1	$ 500	$3,000	×	33.3%	$ 999
2	1,000	3,000	×	44.5	1,335
3	1,000	3,000	×	14.8	444
4	500	3,000	×	7.4	222
	$3,000				$3,000

Example 12.14 — A machine costs $1,000. Annual cash inflows are expected to be $500. The machine will be depreciated using the MACRS rule and will fall under the 3-year property class. The cost of capital after taxes is 10%. The estimated life of the machine is 4 years. The tax rate is 30%. The formula for computation of after-tax cash inflows (S – E)(1 – t)+ (d)(t) needs to be computed separately. The NPV analysis can be performed as follows:

						Present Value Factor @ 10%	Present Value
(S – E)(1 – t): $500(1 – 0.3) = $350 for 4 years						3.170[a]	$1,109.50
Year	**Cost**		**MACRS%**	**d**	**(d)(t)**		
1	$1,000	×	33.3%	$333	$99.9	.909[b]	90.81
2	$1,000	×	44.5	445	133.5	.826[b]	110.27
3	$1,000	×	14.8	148	44.4	.751[b]	33.34
4	$1,000	×	7.4	74	22.2	.683[b]	15.16
							$1,359.08

[a] T_4(10%, 4 years) = 3.170 (from Table 11.4).

[b] T_3 values obtained from Table 11.3.

Therefore, NPV = PV – I = $1,359.08 – $1,000 = $359.08, which is positive, so that the machine should be bought.

12.8 WHAT TO KNOW ABOUT THE COST OF CAPITAL

The cost of capital is defined as the rate of return that is necessary to maintain the market value of the firm (or price of the firm's stock). Project managers must know the cost of capital, often called the *minimum required rate of return*, was used either as a discount rate under the NPV method or as a hurdle rate under the IRR method earlier in the chapter and in calculating the residual income (RI) in Chapter 13. The cost of capital is computed as a weighted average of the various capital components, which are items on the right-hand side of the balance sheet such as debt, preferred stock, common stock, and retained earnings.

12.8.1 Cost of Debt and Preferred Stock

The cost of debt is stated on an after-tax basis, since the interest on the debt is tax deductible. However, the cost of preferred stock is the stated annual dividend rate. This rate is not adjusted for income taxes because the preferred dividend, unlike debt interest, is not a deductible expense in computing corporate income taxes.

Example 12.15 — Assume that the Hume Company issues a $1,000, 8%, 20-year bond whose net proceeds are $940. The tax rate is 40%. Then, the after-tax cost of debt is:

$$8.00\% \, (1 - 0.4) = 4.8\%$$

Example 12.16 — Suppose that the Hume company has preferred stock that pays a $12 dividend per share and sells for $100 per share in the market. Then the cost of preferred stock is:

$$\frac{\text{Dividend per share}}{\text{Price per share}} = \frac{\$12}{\$100} = 12\%$$

12.8.2 Cost of Common Stock

The cost of common stock is generally viewed as the rate of return investors require on a firm's common stock. One way to measure the cost of common stock is to use the *Gordon's growth model*. The model is

$$P_o = \frac{D_1}{r - g}$$

where

P_o = value (or market price) of common stock
D_1 = dividend to be received in 1 year
r = investor's required rate of return
g = rate of growth (assumed to be constant over time)

Solving the model for r results in the formula for the cost of common stock:

$$r = \frac{D_1}{P_o} + g$$

Example 12.17 — Assume that the market price of the Hume Company's stock is $40. The dividend to be paid at the end of the coming year is $4 per share and is expected to grow at a constant annual rate of 6%. Then the cost of this common stock is:

$$\frac{D_1}{P_o} + g = \frac{\$4}{\$40} + 6\% = 16\%$$

12.8.3 Cost of Retained Earnings

The cost of retained earnings is closely related to the cost of existing common stock, since the cost of equity obtained by retained earnings is the same as the rate of return investors require on the firm's common stock.

12.8.4 Measuring the Overall Cost of Capital

The firm's overall cost of capital is the weighted average of the individual capital costs, with the weights being the proportions of each type of capital used, that is,

$\sum$ (percentage of the total capital structure supplied by each source of capital × cost of capital for each source).

The computation of overall cost of capital is illustrated in the following example.

Example 12.18 — Assume that the capital structure at the latest statement date is indicative of the proportions of financing that the company intends to use over time:

		Cost
Mortgage bonds ($1,000 par)	$20,000,000	4.80%
Preferred stock ($100 par)	5,000,000	12.00
Common stock ($40 par)	20,000,000	16.00
Retained earnings	5,000,000	16.00
Total	$50,000,000	

These proportions would be applied to the assumed individual explicit after-tax costs below:

Source	Weights	Cost	Weighted Cost
Debt	40%[a]	4.80%	1.92%[b]
Preferred stock	10	12.00%	1.20
Common stock	40	16.00%	6.40
Retained earnings	10	16.00%	1.60
	100%		11.12%

[a] $20,000,000/$50,000,000 = .40 = 40%.

[b] 4.80% × 40% = 1.92%.

Overall cost of capital is 11.12%.

By computing a company's cost of capital, we can determine its minimum rate of return, which is used as the discount rate in present value calculations and in calculating an investment center's residual income (RI). A company's cost of capital is also an indicator of risk. For example, if your company's cost of financing increases, it is being viewed as more risky by investors and creditors, who are demanding higher return on their investments in the form of higher dividend and interest rates.

12.9 CONCLUSION

We have examined the process of evaluating investment projects. We have also discussed five commonly used criteria for evaluating capital budgeting projects, including the net present value (NPV) and internal rate of return (IRR) methods. The problems that arise with mutually exclusive investments and capital rationing were addressed. Since income taxes could make a difference in the accept or reject decision, tax factors must be taken into account in every decision.

Although the traditional depreciation methods still can be used for computing depreciation for book purposes, 1981 saw a new way of computing depreciation deductions for tax purposes. This rule is called the Modified Accelerated Cost Recovery System (MACRS). It was enacted by Congress in 1981 and then modified somewhat under the Tax Reform Act of 1986. We illustrated the use of MACRS, and presented an overview of the traditional depreciation methods. We also covered how to calculate a firm's cost of capital, which is used either as a discount rate under the NPV method or as a hurdle rate under the IRR method earlier in the chapter and in calculating the residual income (RI) in Chapter 13.

13 How to Analyze and Improve Management Performance

The ability to measure managerial performance is essential in controlling operations toward the achievement of organizational goals. As companies grow or their activities become more complex, they attempt to decentralize decision making as much as possible. They do this by restructuring the firm into several divisions and treating each as an independent business. The managers of these subunits or segments are then evaluated on the basis of the effectiveness with which they use the assets entrusted to them.

Perhaps the most widely used single measure of success of an organization and its subunits is the rate of return on investment (ROI). Related is the return to stockholders, known as the return on equity (ROE). In this chapter, you will learn:

- What ROI is
- The basic components of the Du Pont formula and how it can be used for profit improvement
- How ROI can be increased
- How financial leverage affects the stockholder's return

13.1 WHAT IS RETURN ON INVESTMENT (ROI)?

ROI relates net income to invested capital (total assets). ROI provides a standard for evaluating how efficiently management employs the average dollar invested in a firm's assets, whether that dollar came from owners or creditors. Furthermore, a better ROI can also translate directly into a higher return on the stockholders' equity.

ROI is calculated as:

$$\text{ROI} = \frac{\text{Net profit after taxes}}{\text{Total assets}}$$

Example 13.1 — Consider the following financial data:

Total assets =	\$100,000
Net profit after taxes =	18,000

$$\text{Then, ROI} = \frac{\text{Net profit after taxes}}{\text{Total assets}} = \frac{\$18{,}000}{\$100{,}000} = 18\%$$

The problem with this formula is that it only tells you about how a company did and how well it fared in the industry. It has very little value from the standpoint of profit planning.

13.2 WHAT DOES ROI CONSIST OF? — DU PONT FORMULA

ROI can be broken down into two factors — profit margin and asset turnover. In the past, managers have tended to focus only on the profit margin earned and have ignored the turnover of assets. It is important to realize that excessive funds tied up in assets can be just as much of a drag on profitability as excessive expenses. The Du Pont Corporation was the first major company to recognize the importance of looking at both net profit margin and total asset turnover in assessing the performance of an organization. The ROI breakdown, known as the *Du Pont formula*, is expressed as a product of these two factors, as shown below.

$$\text{ROI} = \frac{\text{Net profit after taxes}}{\text{Total assets}} = \frac{\text{Net profit after taxes}}{\text{Sales}} \times \frac{\text{Sales}}{\text{Total assets}}$$

$$= \text{Net profit margin} \times \text{Total asset turnover}$$

The Du Pont formula combines the income statement and balance sheet into this otherwise static measure of performance. Net profit margin is a measure of profitability or operating efficiency. It is the percentage of profit earned on sales. This percentage shows how many cents attach to each dollar of sales. On the other hand, total asset turnover measures how well a company manages its assets. It is the number of times by which the investment in assets turns over each year to generate sales.

The breakdown of ROI is based on the thesis that the profitability of a firm is directly related to management's ability to manage assets and control expenses effectively.

Example 13.2 — Assume the same data as in Example 13.1. Also assume sales of $200,000.

Then, $\text{ROI} = \frac{\text{Net profit after taxes}}{\text{Total assets}} = \frac{\$18,000}{\$100,000} = 18\%$

Alternatively,

$$\text{Net profit margin} = \frac{\text{Net profit after taxes}}{\text{Sales}} = \frac{\$18,000}{\$200,000} = 9\%$$

$$\text{Total asset turnover} = \frac{\text{Sales}}{\text{Total assets}} = \frac{\$200,000}{\$100,000} = 2 \text{ times}$$

Therefore,

$$\text{ROI} = \text{Net profit margin} \times \text{Total asset turnover} = 9\% \times 2 \text{ times} = 18\%$$

The breakdown provides numerous insights to financial managers on how to improve profitability of the company and investment strategy. (Note that net profit margin and total asset turnover are hereafter called margin and turnover, respectively.) Specifically, it has several advantages over the original formula (i.e., net profit after taxes/total assets) for profit planning. They are:

1. The importance of turnover as a key to overall return on investment is emphasized in the breakdown. In fact, turnover is just as important as profit margin in enhancing overall return.
2. The importance of sales is explicitly recognized, which is not in the original formula.
3. The breakdown stresses the possibility of trading one for the other in an attempt to improve a company's overall performance. The margin and turnover complement each other. In other words, a low turnover can be made up by a high margin, and vice versa.

Example 13.3 — The breakdown of ROI into its two components shows that a number of combinations of margin and turnover can yield the same rate of return, as shown below:

	Margin	×	**Turnover**	**= ROI**
(1)	9%	×	2 times	= 18%
(2)	8	×	2.25	= 18
(3)	6	×	3	= 18
(4)	4	×	4.5	= 18
(5)	3	×	6	= 18
(6)	2	×	9	= 18

The margin-turnover relationship and its resulting ROI are depicted in Figure 13.1. As the figure shows, the margin and turnover factors complement each other. Weak margin can be complemented by a strong turnover, and vice versa. It also shows how turnover is an important key to profit making. In effect, these two factors are equally important in overall profit performance.

13.3 ROI AND PROFIT OBJECTIVE

Figure 13.1 can also be looked at as showing six companies that performed equally well (in terms of ROI), but with varying income statements and balance sheets. There is no ROI that is satisfactory for all companies. Sound and successful operation must point toward the optimum combination of profits, sales, and capital employed. The combination will necessarily vary depending upon the nature of the business and the characteristics of the product. An industry with products tailor-made to customers' specifications will have different margins and turnover ratios, compared with industries that mass-produce highly competitive consumer goods. For example, the combination (4) may describe a supermarket operation that inherently works with

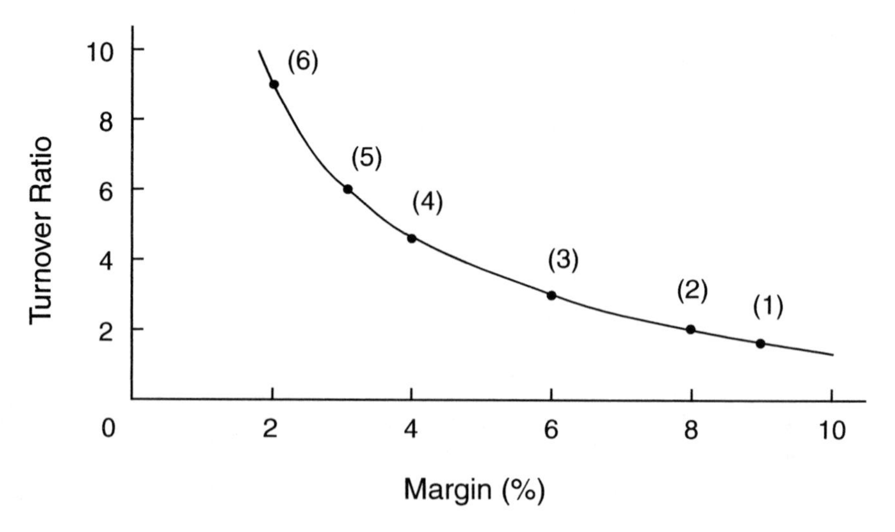

FIGURE 13.1 The margin-turnover relationship.

low margin and high turnover, while the combination (1) may be a jewelry store that typically has a low turnover and high margin.

13.4 ROI AND PROFIT PLANNING

The breakdown of ROI into margin and turnover gives management insight into planning for profit improvement by revealing where weaknesses exist: margin or turnover, or both. Various actions can be taken to enhance ROI. Generally, management can:

1. Improve margin
2. Improve turnover
3. Improve both

Alternative 1 demonstrates a popular way of improving performance. Margins may be increased by reducing expenses, raising selling prices, or increasing sales faster than expenses. Some of the ways to reduce expenses are:

(a) Use less costly inputs of materials, although this can be dangerous in today's quality-oriented environment.
(b) Automate processes as much as possible to increase labor productivity. But this will probably increase assets, thereby reducing turnover.
(c) Bring the discretionary fixed costs under scrutiny, with various programs either curtailed or eliminated. Discretionary fixed costs arise from annual budgeting decisions by management. Examples include advertising, research and development, and management development programs. The cost-benefit analysis is called for in order to justify the budgeted amount of each discretionary program.

A company with pricing power can raise selling prices and retain profitability without losing business. Pricing power is the ability to raise prices even in poor economic times when unit sales volume may be flat and capacity may not be fully utilized. It is also the ability to pass on cost increases to consumers without attracting domestic and import competition, political opposition, regulation, new entrants, or threats of product substitution. The company with pricing power must have a unique economic position. Companies that offer unique, high-quality goods and services (where the service is more important than the cost) have this economic position.

Alternative 2 may be achieved by increasing sales while holding the investment in assets relatively constant, or by reducing assets. Some of the strategies to reduce assets are:

(a) Dispose of obsolete and redundant inventory. The computer has been extremely helpful in this regard, making continuous monitoring of inventory more feasible for better control.
(b) Devise various methods of speeding up the collection of receivables and also evaluate credit terms and policies.
(c) See if there are unused fixed assets.
(d) Use the converted assets (primarily cash) obtained from the use of the previous methods to repay outstanding debts or repurchase outstanding issues of stock. You may use those funds elsewhere to get more profit, which will improve margin as well as turnover.

Alternative 3 may be achieved by increasing sales or by any combinations of alternatives 1 and 2.

Figure 13.2 shows complete details of the relationship of ROI to the underlying ratios — margin and turnover — and their components. This will help identify more detailed strategies to improve margin, turnover, or both.

EXAMPLE 13.4 — Assume that management sets a 20% ROI as a profit target. It is currently making an 18% return on its investment.

$$\text{ROI} = \frac{\text{Net profit after taxes}}{\text{Total assets}} = \frac{\text{Net profit after taxes}}{\text{Sales}} \times \frac{\text{Sales}}{\text{Total assets}}$$

Present situation:

$$18\% = \frac{18{,}000}{200{,}000} \times \frac{200{,}000}{100{,}000}$$

The following are illustrative of the strategies which might be used (each strategy is independent of the other).

Alternative 1: Increase the margin while holding turnover constant. Pursuing this strategy would involve leaving selling prices as they are and making every effort to increase efficiency so as to reduce expenses. By doing so, expenses might be reduced by $2,000 without affecting sales and investment to yield a 20% target ROI, as follows:

$$20\% = \frac{20{,}000}{200{,}000} \times \frac{200{,}000}{100{,}000}$$

Alternative 2: Increase turnover by reducing investment in assets while holding net profit and sales constant. Working capital might be reduced or some land might be sold, reducing investment in assets by $10,000 without affecting sales and net income to yield the 20% target ROI as follows:

$$20\% = \frac{18{,}000}{200{,}000} \times \frac{200{,}000}{90{,}000}$$

Alternative 3: Increase both margin and turnover by disposing of obsolete and redundant inventories or through an active advertising campaign. For example, trimming down $5,000 worth of investment in inventories would also reduce the inventory holding charge by $1,000. This strategy would increase ROI to 20%.

$$20\% = \frac{19{,}000}{200{,}000} \times \frac{200{,}000}{95{,}000}$$

Excessive investment in assets is just as much of a drag on profitability as excessive expenses. In this case, cutting unnecessary inventories also helps cut down expenses of carrying those inventories, so that both margin and turnover are improved at the same time. In practice, alternative 3 is much more common than alternative 1 or 2.

13.5 ROI AND RETURN ON EQUITY (ROE)

Generally, a better management performance (i.e., a high or above-average ROI) produces a higher return to equity holders. However, even a poorly managed company that suffers from a below-average performance can generate an above-average return on the stockholders' equity, simply called the return on equity (ROE). This is because borrowed funds can magnify the returns a company's profits represent to its stockholders.

Another version of the Du Pont formula, called the modified Du Pont formula, reflects this effect. The formula ties together the ROI and the degree of financial leverage (use of borrowed funds). The financial leverage is measured by the equity multiplier, which is the ratio of a company's total asset base to its equity investment, or, stated another way, the ratio of how many dollars of assets held per dollar of stockholders' equity. It is calculated by dividing total assets by stockholders' equity. This measurement gives an indication of how much of a company's assets are financed by stockholders' equity and how much are financed with borrowed funds.

The return on equity (ROE) is calculated as:

$$\text{ROE} = \frac{\text{Net profit after taxes}}{\text{Stockholders' equity}} = \frac{\text{Net profit after taxes}}{\text{Total assets}} \times \frac{\text{Total assets}}{\text{Stockholders' equity}}$$

$$= \text{ROI} \times \text{Equity multiplier}$$

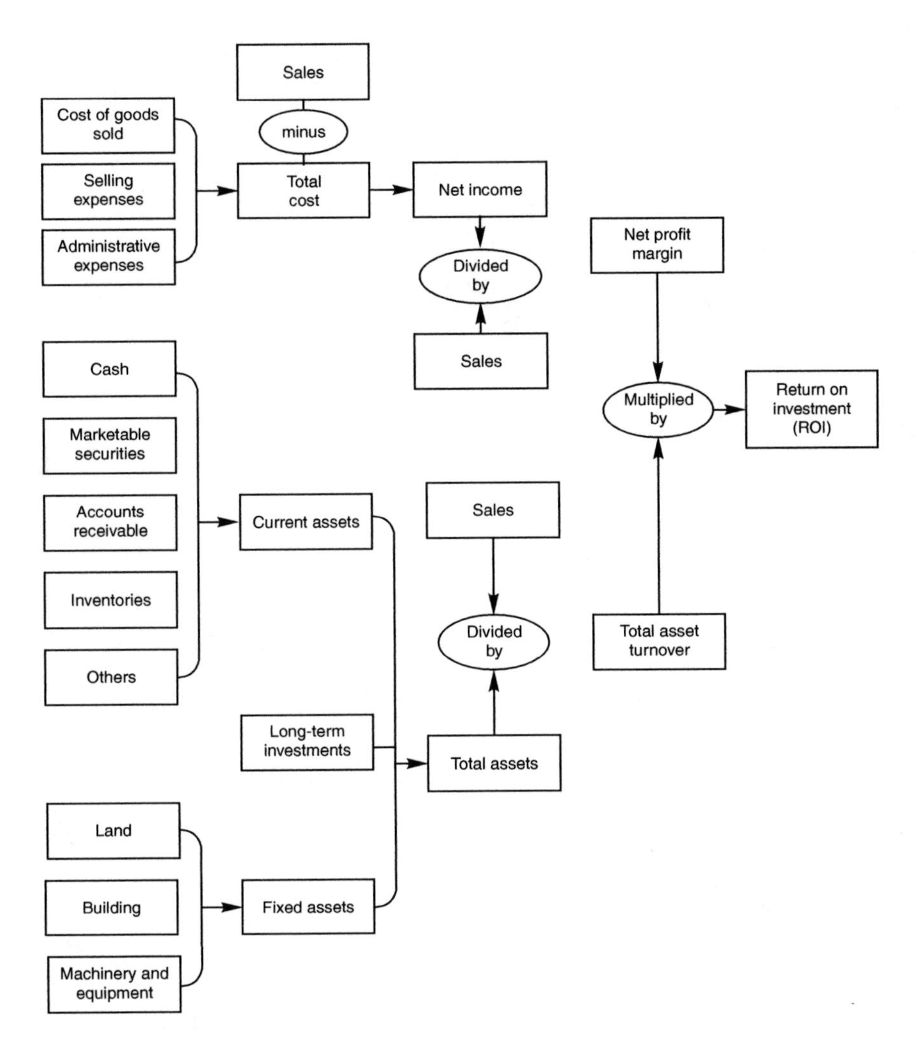

FIGURE 13.2 Relationships of factors influencing ROI.

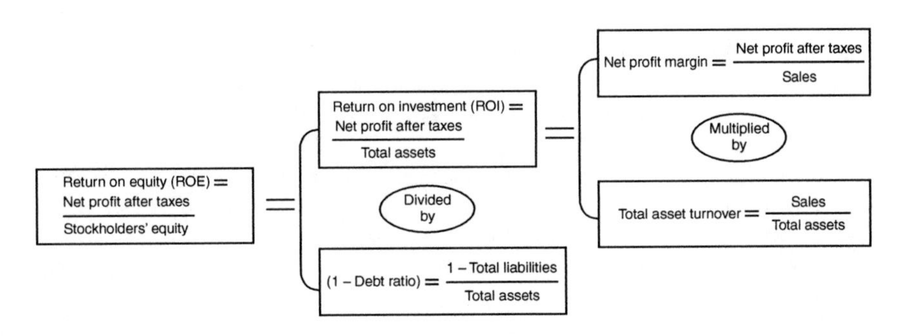

FIGURE 13.3 ROI, ROE, and financial leverage.

ROE measures the returns earned on the owners' (both preferred and common stockholders') investment. The use of the equity multiplier to convert the ROI to the ROE reflects the impact of the leverage (use of debt) on the stockholders' return.

$$\text{The equity multiplier} = \frac{\text{Total assets}}{\text{Stockholders' equity}}$$

$$= \frac{\text{Total assets}}{\text{Total assets} - \text{Total liabilities}}$$

$$= \frac{1}{1 - \dfrac{\text{Total liabilities}}{\text{Total assets}}}$$

$$= \frac{1}{(1 - \text{Debt ratio})}$$

Figure 13.3 shows the relationship among ROI, ROE, and financial leverage.

Example 13.5 — In Example 13.1, assume stockholders' equity of $45,000.

$$\text{Then, Equity multiplier} = \frac{\text{Total assets}}{\text{Stockholders' equity}} = \frac{\$100{,}000}{\$45{,}000} = 2.22$$

$$= \frac{1}{(1 - \text{debt ratio})} = \frac{1}{(1 - 0.55)} = \frac{1}{0.45} = 2.22$$

$$\text{ROE} = \frac{\text{Net profit after taxes}}{\text{Stockholders' equity}} = \frac{\$18{,}000}{\$45{,}000} = 40\%$$

$$\text{ROE} = \text{ROI} \times \text{Equity multiplier} = 18\% \times 2.22 = 40\%$$

If the company used only equity, the 18% ROI would equal ROE. However, 55% of the firm's capital is supplied by creditors ($45,000/$100,000 = 45% is the equity-to-asset ratio; $55,000/$100,000 = 55% is the debt ratio). Since the 18% ROI all goes to stockholders, who put up only 45% of the capital, the ROE is higher than 18%. This example indicates the company was using leverage (debt) favorably.

Example 13.6 — To further demonstrate the interrelationship between a firm's financial structure and the return it generates on the stockholders' investments, let us compare two firms that generate $300,000 in operating income. Both firms employ $800,000 in total assets, but they have different capital structures. One firm employs no debt, whereas the other uses $400,000 in borrowed funds. The comparative capital structures are shown as:

	A	B
Total assets	$800,000	$800,000
Total liabilities	—	400,000
Stockholders' equity (a)	800,000	400,000
Total liabilities and stockholders' equity	$800,000	$800,000

Firm B pays 10% interest for borrowed funds. The comparative income statements and ROEs for firms A and B would look as follows:

	A	B
Operating income	$300,000	$300,000
Interest expense		(40,000)
Profit before taxes	$300,000	$260,000
Taxes (30% assumed)	(90,000)	(78,000)
Net profit after taxes (b)	$210,000	$182,000
ROE [(b)/(a)]	26.25%	45.5%

The absence of debt allows firm A to register higher profits after taxes. Yet the owners in firm B enjoy a significantly higher return on their investments. This provides an important view of the positive contribution debt can make to a business, but within a certain limit. Too much debt can increase the firm's financial risk and thus the cost of financing.

If the assets in which the funds are invested are able to earn a return greater than the fixed rate of return required by the creditors, the leverage is positive and the common stockholders benefit. The advantage of this formula is that it enables the company to break its ROE into a profit margin portion (net profit margin), an efficiency-of-asset-utilization portion (total asset turnover), and a use-of-leverage portion (equity multiplier). It shows that the company can raise shareholder return by employing leverage — taking on larger amounts of debt to help finance growth.

Since financial leverage affects net profit margin through the added interest costs, management must look at the various pieces of this ROE equation, within the context of the whole, to earn the highest return for stockholders. Financial managers have the task of determining just what combination of asset return and leverage will work best in its competitive environment. Most companies try to keep at least a level equal to what is considered to be "normal" within the industry.

13.6 A WORD OF CAUTION

Unfortunately, leverage is a double-edged sword. If assets are unable to earn a high enough rate to cover fixed finance charges, then the stockholder suffers. The reason is that part of the profits from the assets which the stockholder has provided to the firm will have to go to make up the shortfall to the long-term creditors, and he/she will be left with a smaller return than would otherwise have been earned.

13.7 CONCLUSION

This chapter covered in detail various strategies to increase the return on investment (ROI). The breakdown of ROI into margin and turnover, popularly known as the Du Pont formula, provides much insight into: (a) the strengths and weaknesses of a business and its segments, and (b) what needs to be done in order to improve performance. Another version of the Du Pont formula — the modified Du Pont formula — relates ROI to ROE (stockholders' return) through financial leverage. It shows how leverage can work favorably for the owners of the company.

14 How to Evaluate Your Segment's Performance

In the previous chapter, we discussed how to analyze and evaluate managerial performance. We touched on a broad measure of performance — that is, return on investment (ROI). In this chapter, we will focus on the performance of a *segment* within a firm. The ability to measure performance is essential in developing management incentives and controlling the operation toward the achievement of organizational goals. You should know the financial strengths and weaknesses for your responsibility center, and what actions you should take, if any, to improve performance.

A segment is a part or activity of a company for which a manager desires cost or revenue data. Examples of segments are divisions, sales territories, individual stores, service centers, manufacturing plants, sales departments, product lines, distribution channels, processes, programs, geographic areas, types of customers, jobs, and contracts. Segmental reports may be prepared for activity at different levels within your responsibility center and in varying formats, depending on your needs.

Segmental reporting reveals your department's performance. It also shows whether or not your product lines are profitable. Segment reports will help you determine what types of goods are being bought by your customers, what profit you are earning from each customer, which sales territories have a poor sales mix, whether or not your salespeople are doing a good job, and whether or not production workers are performing effectively.

Analysis of segmental performance helps you evaluate the success or failure of your segment. For example, divisional performance measures are concerned with the contribution of the division to profit and quality, as well as whether or not the division meets overall goals.

In evaluating a product line, consideration should be given to profitability, growth, competition, and capital employed, You will have to determine whether to drop unprofitable products or substantially raise prices.

Within a customer class, there may also be a difference in selling costs to different customers within that class. Why? Perhaps some customers need more extensive services (e.g., delivery, warehousing). Customer analysis will show the difference in profitability among customers so that corrective action may be taken. The analysis will aid in formulating selling prices and monitoring and controlling distribution costs. It will help you determine the impact on profitability of proposed price and volume changes.

14.1 APPRAISING MANAGER PERFORMANCE

In appraising your performance as the manager of the segment, you must determine which factors were under your control (e.g., advertising budget) and which factors were not (e.g., economic conditions). Comparison should be made of the performance of your division to other divisions in the company, as well as to similar divisions in competing companies. Appraisal should also be made of the risk and earning potential of your division.

The reasons to measure your performance as a division manager are as follows:

- It assists in formulating incentives and controlling operations to meet company and departmental goals.
- It points to trouble spots needing attention.
- It helps you determine who should be rewarded for good performance.
- It helps you determine who is not doing well so that corrective action may be taken.
- It aids in allocating time among projects.
- It provides job satisfaction since you receive feedback.

Divisional performance is analyzed by a responsibility center, which is composed of a revenue center, a cost center, a profit center, and an investment center.

14.2 RESPONSIBILITY CENTER

Responsibility accounting is the system for collecting and reporting revenue and cost information by areas of responsibility. It operates on the premise that you should be held responsible for your performance, the performance of your subordinates, and for all activities within your responsibility center. It is both a planning and control technique. Responsibility accounting: (1) facilitates delegation of decision making; (2) helps promote "management by objective," in which managers agree on a set of goals (your performance is evaluated based on the attainment of these goals); and (3) permits effective use of "management by exception."

Figure 14.1 shows responsibility centers within an organization, while Figure 14.2 presents an organization chart of a company.

A *responsibility center* is a unit that has control over costs, revenues, and investment funds. This center may be responsible for all three functions or for only one function. Responsibility centers are found in both centralized and decentralized organizations. A profit center is often associated with a decentralized organization, while a cost center is usually associated with a centralized one.

There are lines of responsibility within a company. Shell, for example, is organized primarily by business functions: exploitation, refining, and marketing. General Foods, on the other hand, is organized by product lines. To understand these lines, you should know how your company is organized.

14.2.1 REVENUE CENTER

As the manager of a revenue center, you are responsible for obtaining a target level of sales revenue. An example is a district sales office. The performance report contains the budgeted and actual sales for the center by product, including evaluation.

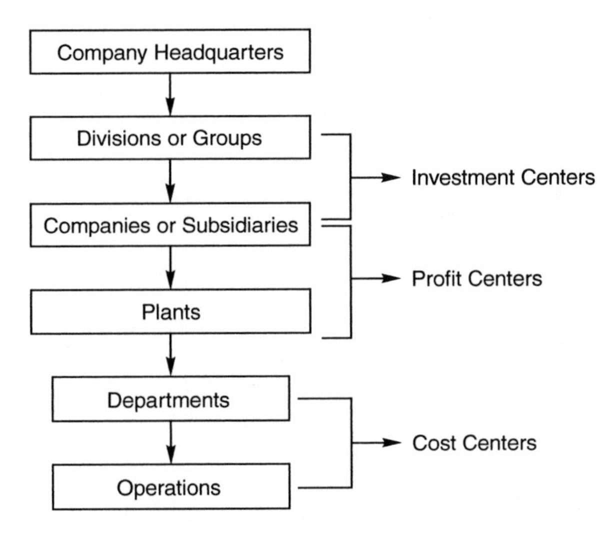

FIGURE 14.1 Responsibility centers within a company.

Usually, you are responsible for marketing a product line. But a revenue center typically has a few costs (e.g., salaries, rent). Hence, you are responsible mostly for revenues and only incidentally for some costs (typically *not* product costs). Accountability for departmental sales revenue also assumes that you have the authority to determine product sales prices.

If you are to generate profitable sales, you must know which areas have the greatest profitability. This requires sound sales analysis and cost evaluation. Sales analysis may involve one or more of the following: prior sales performance, looking at sales trends over the years, and comparing actual sales to budgeted sales. Analyses may be in dollars and/or units. The types of analysis include those by salesperson, terms of sale, order size, channel of distribution, customer, product, and territory. Subanalyses may also be made for evaluating product sales in each territory.

Sales analyses can help you uncover unwanted situations, such as when most of your division's sales are derived from a small share of your product line. Further, you may see that only a few customers give you most of your sales volume. Thus, a small part of the selling effort is needed for a high percentage of your business. After studying the information, you may more narrowly focus your sales effort, resulting in fewer selling expenses. Territorial assignments may be changed, or the product line may be simplified.

Sales analysis will also allow you to see if sales effort is directed toward the wrong products and if sales of products are going toward less profitable lines. In appraising sales effort, consideration should be given to the number of calls, ads placed, and mailings, and their results.

You should keep a record of sales levels for the same product at varying prices so as to establish a relationship of volume to price.

Also, sales orders and back orders may be used as measures of projected volume. You can determine whether orders have been lost by stockouts or delayed shipments.

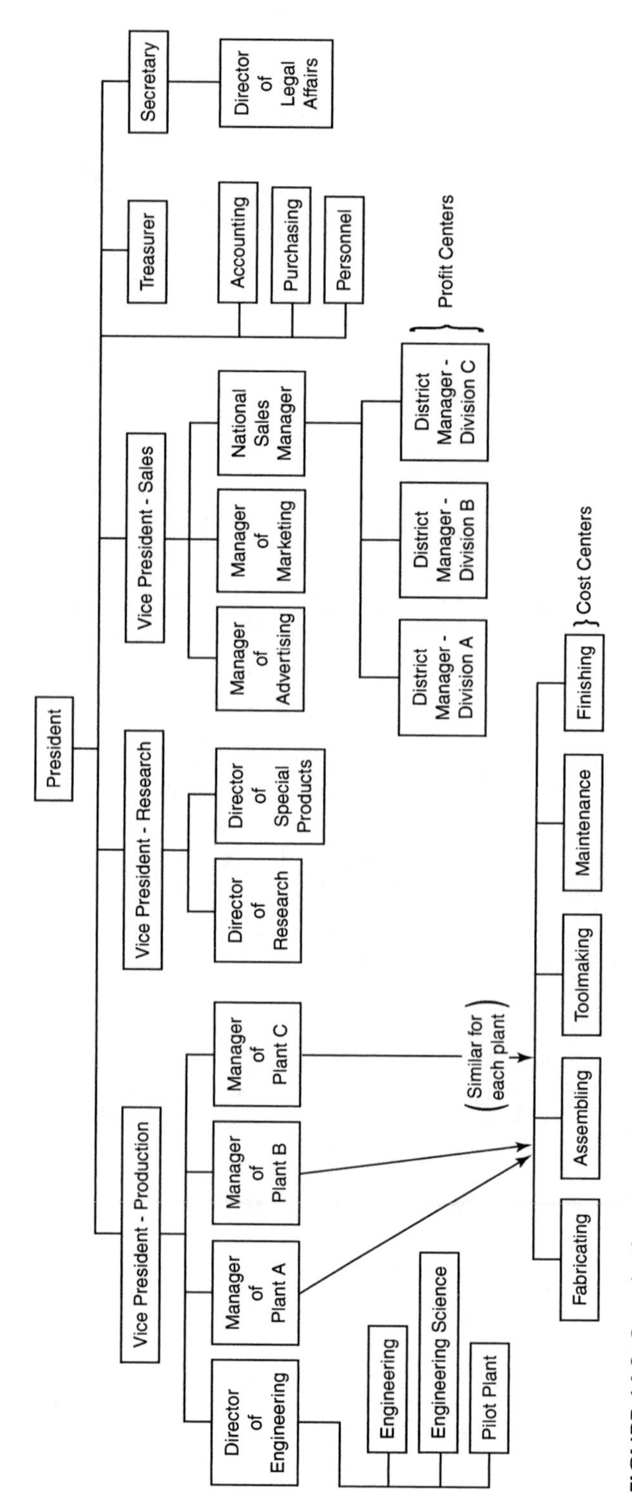

FIGURE 14.2 Organization chart of a company.

14.2.2 Cost Center

A cost center is typically the smallest segment of activity or responsibility for cost accumulation. Examples of cost centers include the maintenance department or the fabricating department in a manufacturing company. Cost centers may be relatively small — a single department with few employees — or very large, such as an administrative area of a large company or an entire factory. Some cost centers may be composed of smaller cost centers. For instance, a factory may be segmented into numerous departments, each of which is a cost center. Generally, cost centers pertain more to departments than to divisions.

A cost center manager is responsible for direct operational costs and for meeting production budgets and quotas. The cost center manager, usually a department head, carries the authority and responsibility for costs and for quantity and quality of not only products but also services. For example, the personnel manager oversees both the costs and the quality of services rendered. Managers of cost centers, however, have no control over sales or marketing activities. Departmental profit is difficult to derive because of problems in allocating revenue and costs.

With a cost center, you must compare budgeted cost to actual cost. Variances are investigated to determine the reasons for particular costs; necessary corrective action is taken to correct problems; and efficiencies are accorded recognition. The cost center approach is useful when you possess control over your costs at a specific operating level.

You may use the cost center approach when problems arise in relating financial measures to output. Cost center evaluation is most suitable for the following functions, where problems in quantifying the output in financial terms occur: accounting and financial reporting, legal, computer services, marketing, personnel, and public relations.

You will also find the cost center approach appropriate for nonprofit and governmental units where budgetary appropriations are assigned. Actual expenditures are compared to budget. Your performance depends on your ability to achieve output levels given budgetary constraints.

When looking at your performance, say at bonus time, the relevant costs are those incremental costs you have control over. Incremental costs are those expenditures that would not exist if the center were abandoned. Hence, allocated common costs (e.g., general administration) should not be included in appraising your performance. Such costs should, however, be allocated in determining the profit for the division. Cost allocation must conform to department goals and should be applied consistently among divisions.

Cost center evaluation will not be worthwhile unless reliable budget figures exist. If a division's situation significantly changes, an adjustment to the initial budget is necessary. In such a case, actual cost should be compared with the initial budget figure (original goal) and the revised budget. Flexible budgets should be prepared to enable you to look at costs at different levels of capacity. For example, you can budget figures for expected capacity, optimistic capacity, and pessimistic capacity. Better still, comparisons of budget to actual cost can thus be made, given changing circumstances.

If you are a manager in the factory, you should consider factory performance measures such as yield percentages for direct materials, number of rejects, capacity utilization, and cost per labor hour.

Provision should exist for chargebacks, where appropriate. For example, if the quality control department makes an effort in its evaluation which leads to acceptance of the product by the purchasing department, the former should be charged with the increased costs necessary to meet acceptable standards incurred by the purchasing department.

When a transfer occurs between cost centers, the transfer price should be based on either actual cost, standard cost, or controllable cost. Transfer price is the price charged between divisions for a product or service. *Warning*: Using actual cost can pass cost inefficiencies on to the next division. There is no incentive to control costs. *Solution*: Using standard cost affects this problem because the selling division will only be credited for what the item *should* cost.

A good transfer price is a controllable cost. The cost center should be charged with actual controllable cost and credited with standard controllable cost for the assembled product or service passed to other divisions. By including just controllable cost, the subjectivity of the allocation of fixed noncontrollable cost does not exist.

You should be aware of the effects of discretionary costs when evaluating performance. Discretionary costs are those that can be easily changed, such as advertising and research. A cutback in discretionary costs will prompt short-term improvements in profitability but in the long run will likely have a negative effect.

In evaluating administrative functions, performance reports should examine such dollar indicators as executive salaries and service departments as well as nondollar measures such as number of files handled, phone calls taken, and invoices processed.

14.2.3 Profit Center

Profits can be expressed in several ways including net income, margin, gross profit, controllable profit, and incremental profit. A center thus measures the performance of a division, product line, graphic area, or any other quantifiable unit, using one of these expressions, On the other hand, profit centers have associated revenues and expenses; net income and contribution margin can be computed for them. However, profit centers may not have significant amounts of invested capital.

A typical profit center is a division selling limited numbers of products or serving a particular geographic area. The profit center provides not only goods and services but also the means for marketing them.

In some instances, profit centers are formed when products or services are used solely within the company. For example, the computer department may be considered a profit center, as it bills each of the administrative and operating units for services rendered.

A profit center demonstrates the following characteristics: (1) defined profit objective; (2) managerial authority for making decisions that have an impact on earnings; and (3) the use of profit-oriented decision rules.

The profit center approach enhances decentralization and delineates units for decision-making purposes. It should be used if your division is self-contained (with its own manufacturing and distribution facilities) and when there is a limited number of interdivisional transfers. The reason is that the profit reported by the

TABLE 14.1
Contribution Margin Income Statement By Segments

		Divisional Breakdown		Breakdown of Division Y				
	Entire Company	Division X	Division Y	Unallocable	Product 1	Product 2	Product 3	Product 4
Sales	$1,200	$300	$900		$300	$100	$200	$300
Less: Variable manufacturing cost of sales	(700)	(200)	(500)		(100)	(50)	(100)	(250)
Manufacturing contribution margin	500	100	400		200	50	100	50
Less: Variable selling and administrative costs	(200)	(50)	(150)		(40)	(40)	(40)	(30)
Contribution margin	300	50	250		160	10	60	20
Less: Controllable fixed costs by segment managers	(180)	(40)	(140)	(40)[a]	(30)	(5)	(50)	(15)
Contribution controllable by segment managers	120	10	110	(40)	130	5	10	5
Less: Fixed costs controllable by others	(70)	(6)	(64)	(20)	(20)	(15)	(5)	(4)
Segmental contribution	50	4	46	(60)	110	(10)	5	1
Less: Unallocated costs	(23)							
Net income	$27							

[a] Only those costs logically traceable to a product line should be allocated.

division is basically independent of other divisions' operating activities, performance efficiencies, and managerial decisions. Further, divisional earnings should not be increased by any action reducing overall corporate profitability. Also, use the profit center approach when you have decision-making authority in terms of the quantity and mix of goods or services. Because, with a profit center, net income is determined as if the division were a separate economic entity, you should be more cognizant of outside market considerations.

In using the profit center approach, it is not essential to allocate fixed costs. In general, fixed costs are much less controllable than variable costs, which are common to a group of divisions. Hence, contribution margin may be a good indicator of your division's performance because it emphasizes cost behavior patterns and controllability of costs that are generally useful for evaluating performance and making decisions. The contribution margin approach aids in computing selling price, output levels, the price to accept an order given an idle capacity situation, maximization of resource uses, and break-even analysis. Ultimately, if your division meets its target contribution margin, any excess will be adequate to cover general corporate expenses.

To evaluate divisional and managerial performance, a contribution income statement can be prepared. Table 14.1 presents a contribution margin income statement by division, with a further breakdown into product lines. Table 14.2 illustrates product line profitability.

TABLE 14.2
Contribution by Products

	Entire Company	Product A	Product B
Projected sales	$100,000	$60,000	$40,000
Variable costs			
Goods sold	30,000	10,000	20,000
Marketing	5,000	4,000	1,000
Less: Total variable costs	(35,000)	(14,000)	(21,000)
Contribution margin	65,000	46,000	19,000
Direct fixed costs			
Production	4,000	2,000	2,000
Marketing	3,000	2,000	1,000
Less: Total direct fixed costs	(7,000)	(4,000)	(3,000)
Profit contribution	58,000	42,000	16,000
Common fixed costs			
Production	10,000		
Marketing	8,000		
Administrative and general	5,000		
Less: Total common costs	(23,000)		
Income before tax	$35,000		

A difficulty with the profit center approach is that profit is calculated after subtracting noncontrollable costs or costs not directly related to divisional activity that have been arbitrarily allocated. The ensuing profit figure may be erroneous. However, cost allocation is required, since divisions must incorporate nondivisional costs that have to be met before the company will show a profit. Policies optimizing divisional earnings will likewise optimize corporate earnings even before the allocation of nondivisional expenses.

It is important to recognize that, while an uncontrollable income statement item is included in appraising the performance of a profit center, it should not be used in evaluating you. An example is the effect of foreign exchange transaction gains and losses because you may have no control over it.

As a profit center *manager*, you are responsible for not only profit and loss items attributable directly in your division but also costs incurred outside of the center (e.g., headquarters, other divisions) for which your center will be billed directly.

Advantages of the profit center approach are that it creates competition in a decentralized company, provides goal congruence between a division and the company, and aids performance evaluation. A drawback is that profits can be manipulated since expenses may be shifted among periods. Examples of discretionary costs where management has wide latitude are research and repairs. Also, this approach does not consider the total assets employed in the division to obtain the profit.

Example 14.1 — It is important to know at what point to sell an item in order to maximize profitability. You can sell a product at its intermediate point in Division A for $170 or its final point in Division B at $260. The outlay cost in Division A is $120, while the outlay cost in Division B is $110. Unlimited product demand exists for both the intermediate product and the final product. Capacity is interchangeable. Divisional performance follows:

	Division A	Division B
Selling Price	$178	$260
Less: Outlay cost A	(120)	(120)
Outlay cost B		(110)
Profit	$ 50	$ 30

Sell at the intermediate point because of the higher profit.

Other measures in appraising divisional performance that are not of a profit nature but that must be considered are the following:

- Ratios between cost elements and assets to appraise effectiveness and efficiency.
- Productivity measures, including input-output relationships. An example is labor hours in a production run. You have to consider the input in time and money, and the resulting output in quantity and quality. Does the maintenance of equipment ensure future growth?
- Personnel development (e.g., number of promotions, turnover).
- Market measures (e.g., market share, product leadership, growth rate, customer service).
- Product leadership indicators (e.g., patented products, innovative technology, product quality, safety record).
- Human resource relationships (e.g., employee turnover rate, customer relations, including on-time deliveries).
- Social responsibility measures (e.g., consumer medals).

14.2.3.1 Transfer Pricing

The transfer price is the one credited to the selling division and charged to the buying division for an internal transfer of an assembled product or service. The transfer resembles, for each division, an "arm's length" transaction. A transfer price has to be formulated so that a realistic and meaningful figure can be determined for your division. It should be established after proper planning. Thus, you need to know what monetary values and market prices to assign to these exchanges or transfers — some version of either. Unfortunately, there is no single transfer price that will please everybody — that is, top management, the selling division, and the buying division — involved in the transfer.

Transfer prices are not only important in performance, but also in decisions involving whether to make or buy, whether to sell or process alternative production possibilities. Further, understanding the transfer pricing can help you account for possible cost overruns. You need to know what transfer price is being used for transfers in a department and why it is being used.

The best transfer price is the negotiated market value of the assembled product or service since it is a fair price and treats each profit center as a separate economic entity. The negotiated market value equals the outside service fee or selling price for the item (a quoted price for a product or service is only comparable if the credit terms, grad, equality, delivery, and auxiliary conditions are precisely the same) less internal cost savings that result from dealing within the organization (e.g., advertising, sales commission, delivery charges, credit, and collections costs). The market value of services is based on the going rate for a specific job (e.g., equipment tune-up) and/or the standard hourly rate (e.g., the hourly rate for an engineer). Market price may be determined from price catalogues, outside bids, and published data on complete market transactions. If two divisions cannot agree on the transfer price, it may be settled by arbitration at a higher level. A temporarily low transfer price (e.g., due to oversupply of the item) or high transfer price (e.g., due to a strike situation causing a supply shortage) should not be employed. An average long-term market price should be used.

A negotiated transfer price works best when outside markets for the intermediate product exist, all parties have access to market information, and you are permitted to deal externally if a negotiated settlement is impossible. If one of these conditions is violated, the negotiated price may break down and cause inefficiencies.

The negotiated market price can tell you whether an item should be produced internally or bought through an outside vendor. If the cost of an item produced internally is more than the market price, you may want to go to your superiors and suggest that the item no longer be supplied internally. You can suggest that the company buy the item through a vendor, saving the company money overall. If the item is a by-product, this could tell management that the cost of the materials is too high. This could make the purchasing department look for another vendor. If the negotiated market value is too high, it could also show that the selling department is wasting money.

If the outside market price is inappropriate or not ascertainable (e.g., new product, absence of replacement market, or too costly to be used for transfer pricing), you should use budgeted cost plus profit markup because this transfer price approximates market value and will highlight divisional inefficiencies. For example, if budgeted cost is $10 and a profit markup on a cost of 20% is desired, the transfer price will be $12. Profit market should take into account the particular characteristics of your division (e.g., product line) rather than the overall corporate profit margin.

There is an incentive to the selling division to control costs because it will not be credited for an amount exceeding budgeted cost plus a markup. Thus, if the selling division's inefficiencies result in excessive actual costs, it would have to absorb the decline in profit to the extent that actual cost exceeds budgeted cost.

Even though actual cost plus profit markup may be used, it has the drawback of passing on cost inefficiencies. In fact, the selling division is encouraged to be cost-inefficient, since the higher its actual cost is, the higher its selling price (since it shows a greater profit) will be. Some division managers use actual cost as the transfer price because of its ease to use, but the problem is that no profit is shown by the selling division and cost inefficiencies are passed on. Further, the cost-based method treats the divisions as cost centers rather than profit or investment centers. Therefore, measures such as return on investment and residual income cannot be used for evaluation purposes.

A transfer price based on cost may be appropriate when minimal services are provided by your department to another. If other departments provide identical or very similar services, a cost-based transfer price may be used since the receiving department will select the services of the department providing the highest quality. Thus, the providing department has an incentive to do a good job. A nonfinancial manager neatly illustrates this situation:

> The two major areas of transfer pricing that directly affect my department are internal training and computer equipment. Training, although internal, is also offered to the public. Because of this competitive situation, training costs are charged to my department at the same amount as to outside companies. Computer equipment, on the other hand, is sold (or leased) to my department at a discount over the price offered to the general public. This is attributable to the fact that computer sales for the company have been rather sluggish, and the discount offers an incentive to buy the company's brand over a competitor's.

We now see how transfer prices are used to determine divisional profit.

Example 14.2 — Division A manufactures an assembled product that can be sold to outsiders or transferred to Division B. Relevant information for the period follows:

Division A	Units
Production	1,500
Transferred to Division B	1,200
Sold outside	300
Selling price $25	
Unit cost $5	

The units transferred to Division B were processed further at a cost of $7. They were sold outside at $45. Transfers are at market value.

Division profit is computed as follows:

	Division A	Division B	Company
Sales	$ 7,500	$54,000	$61,500
Transfer price	30,000		
	$37,500	$54,000	$61,500
Product cost	$ 7,500	$ 8,400	$15,900
Transfer price		30,000	
	($ 7,500)	$38,400	$15,900
Profit	$30,000	$15,600	$45,600

We now see whether a buying division should buy inside or outside.

Example 14.3 — An assembly division wants to charge a finishing $80 per unit for an internal transfer of 800 units. The variable cost per unit is $50. Total fixed cost in the assembly division is $200,000. Current production is 10,000 units. Idle capacity exists. The finishing division purchases the item outside for $73 per unit.

The maximum transfer price should be $73 because the finishing division should not have to pay a price greater than the outside market price.

Whether the buying division should be permitted to buy the item outside or be forced to buy inside depends on what is best for overall corporate profitability. Typically, the buying division is required to purchase inside at the maximum transfer price ($73), since the selling division still has to meet its fixed cost when idle capacity exists. The impact on corporate profitability of having the buying division go outside is determined as follows:

Savings to assembly division	
[Units (800) × Variable cost per unit ($50)]:	$40,000
Cost to finishing division	
[Units (800) × Outside selling price ($73)]:	58,400
Stay inside savings	$18,400

The buying division will be asked to purchase inside the company because, if it went outside, corporate profitability would decline by $18,400.

A nonfinancial manager summarizes the importance of transfer pricing policy:

The nonfinancial manager needs to know what the transfer pricing policy of the company is. If he thinks he is being overcharged by a department, he should go to the manager of the other department and tell him that he can get the item from another location for a better price. He could tell the other manager either lower the price or he will go to the other location. If the manager can lower the price and believes you are serious, he would probably lower the price for two reasons. One, even though it's an internal department, he still needs the business. Second, if upper management finds out that people are using outside vendors, it could prove embarrassing to the selling department. The selling department does not want upper management to think they are mismanaged. They also do not want the auditing division to check their records. Even if there is nothing to hide, no one wants the auditors in their department.

A manager may be trying to decide whether he should shut down an operation due to the lack of profit. Transfer pricing could help him determine which course of action to take. If the operation produces a material that is used in other departments to create a good product, the manager can determine what the actual cost to produce the material is. If he determines that he can get the material for a lower price from an outside vendor and that vendor can produce the quantity he needs at the time he needs them, the manager can make a strong case for eliminating the operation and using an outside source. This will cut costs and give him a higher profit margin, and make him look good to his superiors.

The stationery department in my company handles the supplies for the various departments. Most of the items in the inventory are special orders. However, there are items that can also be bought in any stationery store. The cost for most of the items in the stationery department is higher than the price the other departments would pay from an outside source. The company has no policy which states that the departments must buy from the stationery department. Most departments will get items from the outside for several reasons. First, the price is lower. There is no incentive for the managers to buy within. Second, it takes several weeks for the merchandise to arrive. If the manager needs the item in a rush the stationery department can do an emergency delivery but the cost is so much, if it's available on the outside, the manager will buy it externally. The stationery department charges the other departments based on their actual costs. This leads me to the conclusion that either the stationery department is getting overcharged from outside vendors or that the department is wasting a lot. Another possibility is that someone in the stationery department is getting a kickback from a vendor. For the specialized products, the other departments have no choice but to use the stationery department.

14.2.4 Investment Center

An investment center has control over revenue, cost, and investment funds. It is a profit center whose performance is evaluated on the basis of the return earned on investment capital. An example is an auto product line that is manufactured on company premises. Investment centers are widely used in highly diversified companies.

A divisional investment is the amount placed in that division and put under your control. Two major divisional performance indicators are *return on investment* (ROI) and *residual income* (RI). Assets assigned to a division include direct assets in the division and allocated corporate assets. Assets are reflected at book value.

You should be able to distinguish between controllable and noncontrollable investment. While the former is helpful in appraising your performance, the latter is used to evaluate the entire division. Controllable investment depends on the degree of your division's autonomy. As an investment center manager, you accept responsibility for both the center's assets and the controllable income.

14.2.4.1 Return on Investment (ROI)

Although ROI was discussed in a prior chapter, we would like to focus on its usefulness as a measure of a manager's performance.

Return on investment equals:

$$\frac{\text{Operating income}}{\text{Operating assets}}$$

An alternative measure is:

$$\frac{\text{Controllable operating profit}}{\text{Controllable net investment (Controllable assets} - \text{Controllable liabilities)}}$$

Example 14.4 computes ROI for a hypothetical situation.

Example 14.4 — The following financial data pertains to your division:

Operating assets	$100,000
Operating income	$ 18,000
Sales	$200,000

Your ROI is:

$$\frac{\text{Operating income}}{\text{Operating assets}} = \frac{\$18,000}{\$100,000} = 18\%$$

The advantages of ROI as a performance indicator are as follows:

- It focuses on maximizing a ratio instead of improving absolute profits.
- It highlights unprofitable divisions.
- It can be used as a base against which to evaluate divisions within the company and to compare the division to a similar division in a competing company.
- It assigns profit responsibility.
- It aids in appraising performance.
- It places emphasis on high-return items.
- It represents a cumulative audit or appraisal of all capital expenditures incurred during your division's existence.
- It is the broadest possible measure of financial performance. Because divisions are often geographically far apart, you are given broad authority in using division assets and acquiring and selling assets.
- It helps make your goals coincide with those of corporate management.
- It focuses on maximizing a ratio instead of improving absolute profits. Alternative profitability measures could be used in the numerator besides net income (e.g., gross profit, contribution margin, segment margin).
- Different assets in the division must see the same return rate, regardless of the asset risk.
- Established rate of return may be too high and could discourage incentive. A labor-intensive division generally has a higher ROI than a capital-intensive one.
- ROI is a static indicator; it does not show future flows.
- A lack of goal congruence may exist between the company and your division. For example, if your company's ROI is 12%, your division's ROI is 18%, and a project's ROI is 16%, you will not accept the project because it will lower your ROI, even though the project may be best for the entire company.
- ROI ignores risk.
- ROI emphasizes short-run performance instead of long-term profitability. To protect the current ROI, you may be motivated to reject other profitable investment opportunities.
- If the projected ROI at the beginning of the year is set unrealistically high, discouragement of investment center incentive could result.

An understanding of ROI enables you to see how your performance is measured and how your department compares with other departments in the organization.

In an interview, a vice-president of a company commented on the usefulness of ROI in evaluating nonfinancial managers:

> There are two department managers in separate areas with a goal of 15% ROI. Manager A decides that to reach the goal he must increase the sales force by 15 agents. It takes six months to complete the training, become fully functional, and contribute premium dollars. The training costs come to $60,000 per agent, so we see that Manager A has invested heavily in human resources to increase income and affect his ROI. Manager B decides to appeal to the existing sales force to increase their efficiency and production with promises of no further hiring and encourages them to push the more profitable property lines. Therefore, Manager B increases his department's ROI by increasing income while holding invested capital steady.
>
> You must evaluate these two managers one year later with regard to the company's goal of 15% ROI and profitable growth. Without a full understanding of the implications and limitations of ROI, you decide that Manager B has achieved his goals, while Manager A has only had his people functional for six months and has not reached his 15% goal. However, being knowledgeable in accounting, you determine that Manager B has merely opted to look good in the short term, but has actually jeopardized the company in the long run by his failure to adequately forecast and plan accordingly. Manager B has not invested wisely in human resources leaving you one year later with a sales force working to capacity with limited growth potential, as they have saturated their markets and are not prepared to invade new markets. Manager A on the other hand has increased his potential by investing wisely and now has 15 additional agents who are fully trained and prepared to lead future company growth by expanding into untapped markets.
>
> Taking a closer look at ROI reveals three added viewpoints. ROI can be positively affected by decreasing costs, increasing revenue, or decreasing total assets. As a nonfinancial manager, you must be fully aware of these factors so that you can evaluate your position accurately as each option presents its own set of consequences. The proper use of accounting allows you to exercise the proper management controls, along with allowing you to plan and monitor your financial position as you move the organization towards its goals.

14.2.4.2 Residual Income (RI)

An alternative measure of divisional performance is residual income, which equals divisional operating income less minimum return times total operating assets.

Example 14.5 — Divisional operating income is $250,000, total operating assets are $2 million, and the cost of capital is 9%. Residual income equals:

Divisional operating income	$250,000
Less: Minimum × total operating assets	
(9% × $2,000,000)	(180,000)
Residual income	$70,000

The minimum return rate is based upon the company's overall financial cost adjusted for your division's risk.

Residual income may be projected by division, center, or specific program to ensure that the rate of return on alternative investments is met or improved on by each segment.

A target residual income may be formulated to act as your objective. The trend in residual income to total available assets should be examined in appraising divisional performance.

The advantages of residual income are as follows:

- The same asset may be required to earn the same return rate irrespective of the division the asset is in.
- Different return rates may be employed for different types of assets, depending on riskiness.
- Different return rates may be assigned to different divisions, depending on the risk of those divisions.
- RI provides an economic income, taking into account the opportunity cost of tying up assets in the division.
- RI identifies operating problem areas.
- RI maximizes dollars instead of a percentage, thereby motivating you to accept all profitable investments.

The disadvantages of residual income include the following:

- Assigning a minimum return involves estimating a risk level that is subjective.
- It may be difficult to determine the valuation basis and means of allocating assets to divisions.
- If book value is used in valuing assets, residual income will artificially increase over time, since the minimum return times total assets becomes lower as the assets become older.
- RI cannot be used to compare divisions of different sizes. Residual income tends to favor the larger divisions due to the large amount of dollars involved.
- Since it is a mixture of controllable and uncontrollable elements, there is no segregation.

14.2.4.3 Decisions Under ROI and RI

The decision whether to use ROI or RI as a measure of your division's performance affects your investment decisions. Under the ROI method, you tend to accept only the investments whose returns exceed the division's ROI. Otherwise, the division's overall ROI would decrease if the investment were accepted. Under the RI method, on the other hand, you would accept an investment as long as it earns a rate in excess of the minimum required rate of return. The addition of such an investment will increase the division's overall RI.

Example 14.6 — The following data apply to your division:

Operating assets	$100,000
Operating income	$18,000
Minimum required rate of return	13%
ROI	18%
RI	$5,000

Assume that you are presented with a project that would yield 15% on a $10,000 investment. You would not accept this project under the ROI approach since the division is already earning 18%. Acquiring this project will bring down the present ROI to 17.73%, as shown below:

	Present	New Project	Overall
Operating assets (a)	$100,000	$10,000	$110,000
Operating income (b)	18,000	1,500*	19,500
ROI (b/a)	18%	15%	17.73%

* $10,000 × 15% = $1,500

Under the RI approach, the manager would accept the new project because it provides a higher rate than the minimum required rate of return (15% vs. 13%). Accepting the new project will increase the overall residual income to $5,200, as shown below:

	Present	New Project	Overall
Operating assets (a)	$100,000	$10,000	$110,000
Operating income (b)	18,000	1,500	19,500
Less: Minimum required income at 13% (c)	(13,000)	(1,300)*	(14,300)
RI (b – c)	$ 5,000	$ 200	$ 5,200

* $10,000 × 13% = $1,300

14.3 CONCLUSION

It is essential to evaluate your segment's performance to identify problem areas. Controllable and uncontrollable factors must be considered. The various means of evaluating performance include cost center, profit center, revenue center, and investment center. Proper analysis under each method is vital in appraising operating efficiency. You should understand the advantages and disadvantages of each method as well as determine which method is most appropriate for a situation.

You should be familiar with the profit and loss statements by territory, commodity, method of sale, customer, and salesperson. The profit and loss figures will indicate areas of strength or weakness.

Product line analysis identifies areas that need corrective action such as changing the selling price, eliminating unprofitable products, emphasizing profitable products, modifying advertising strategy, and selecting a center of distribution.

There are other goals that managers have which include increasing market share (e.g., your division's sales relative to sales of divisions in c companies, number of new customers), improved productivity (e.g., reduction in per unit cost), and improving employee morale (e.g., turn revenue per employee).

If you are in a service business, some performance measures include billable time, average billing rates, and cost per hour of employee time.

15 How Taxes Affect Business Decisions

Since your company must pay taxes, it is important that you have an understanding of the basic concepts underlying the prevailing tax structure and of how these concepts affect your decisions. Federal, state, and local taxes may be levied on income, sales, and property. It is important that you understand the federal corporate tax because it is the largest in magnitude and thus influences your managerial decisions. If your company is operating internationally, tax laws of foreign countries must also be understood.

To make sound financial and investment decisions, you should be familiar with tax strategies, taxable income computations, deductible expenses, capital gains and losses, operating loss carrybacks and carry forwards, tax credits, and S corporations.

15.1 TAX STRATEGIES AND PLANNING

Sound planning minimizes tax obligations and postpones the payment of taxes to later years. However, ignoring taxes will cause an overstatement in estimated income. As a result, an investment alternative may be chosen that does not sufficiently generate the needed return for the risk exposure taken. You must analyze the tax consequences of such alternative approaches. Income and expenses must be shifted into tax years that will result in the least tax, which means you have to keep up to date with the changes in the tax law that affect your business.

Corporate taxes should be deferred when: (1) there will be a lower tax bracket in a future year; (2) the firm lacks the funds to meet the present tax requirement; (3) the business earns a return on the funds for another year that would have had to have been paid to the federal and local taxing authorities; (4) the tax payment will be in "cheaper" dollars; or (5) there may eventually be no tax payment (e.g., as a result of a new tax law).

You should also properly time the receipt of income and the payment of expenses to minimize tax payment. A good tax strategy is to receive income in a year it will be taxed at a lower rate. For example, if you expect tax rates to drop next year, you can reduce the tax obligation by deferring income to next year. Try to convert income to less-taxed sources. Also, pay deductible expenses in a year in which you will receive the most benefit. For example, accelerate deductions in the current year if you anticipate lower rates in the next year and accelerate expenses that will no longer be deducted or that will be restricted in the future.

Another tax strategy your company might consider is to donate appreciated property instead of cash. By donating appreciated property to a charity, business can deduct the full market value and avoid paying tax on the gain. In some cases, a company can use one method of accounting for books and another method for tax. This is referred to as *interperiod tax allocation* because the temporary difference eventually reverses. A company that uses an accounting method that results in less taxable income initially will lower its tax in the earlier years due to the time value of money. For example, taxes can be reduced in the initial years by using the modified accelerated cost recovery system (MACRS) depreciation rather than straight-line depreciation for tangible assets.

Obtaining tax-exempt income is another tax strategy your company should consider. Tax-free income is worth much more than taxable income. You can determine the equivalent taxable return as follows:

$$\text{Equivalent taxable return} = \frac{\text{Tax-free return}}{1 - \text{Marginal tax rate}}$$

Interest earned on municipal bonds, for example, is not subject to federal tax and is exempt from tax of the state in which the bond was issued. Of course, the market value of the bond changes with changes in the *going interest rate*. The capital gain or loss (difference between cost and sell price) is subject to tax when your company sells the bond.

Example 15.1 — A municipal bond pays an interest rate of 6%. The company's tax rate is 34%. The equivalent rate on a taxable instrument is:

$$\frac{.06}{1.34} = \frac{.06}{.66} = 9.1\%$$

15.2 TAX COMPUTATION

Corporations pay federal income tax on their taxable income, which is the corporation's gross income reduced by the deductions permitted under the Internal Revenue Code of 1986. Federal income taxes are imposed at the following tax rates:

15% on the first	$ 50,000
25% on the next	$ 25,000
34% on the next	$ 25,000
39% on the next	$ 235,000
34% on the next	$9,665,000
35% on the next	$5,000,000
38% on the next	$3,333,333
35% on the remaining income	

Example 15.2 — If a firm has $20,000 in taxable income, the tax liability is $3,000 ($20,000 × 15%).

Example 15.3 — If a firm has $20,000,000 in taxable income, the tax is calculated as follows:

Income ×	Marginal Tax Rate (%)	=	Taxes
$ 50,000	15%		$ 7,500
25,000	25%		6,250
25,000	34%		8,500
235,000	39%		91,650
9,665,000	34%		3,286,100
5,000,000	35%		1,750,000
3,333,333	38%		1,266,667
1,666,667	35%		583,333
$20,000,000			$7,000,000

Financial managers often refer to the federal tax rate imposed on the next dollar of income as the "marginal tax rate" of the taxpayer. Because of the fluctuations in the corporate tax rates, financial managers also talk in terms of the *average tax rate* of a corporation. Average tax rates are computed as follows:

Average tax rate = Tax due/taxable income

Example 15.4 — The average tax rate for the corporation in Example 15.3 is 35% (7,000,000/20,000,000). The marginal tax rate for the corporation in Example 15.3 is 35%.

As suggested in Example 15.4, at taxable incomes beyond $18,333,333, corporations pay a tax of 35% on all of their taxable income. This fact demonstrates the reasoning behind the patch-quilt of corporate tax rates. The 15%, 25%, and 34% tax brackets demonstrate the intent that there should be a graduated tax rate for small corporate taxpayers. The effect of the 39% tax bracket is to wipe out the early low tax brackets. At $335,0000 of corporate income, the cumulative income tax is $113,900, which results in an average tax rate of 34% ($113,900/$335,000). The income tax rate increases to 35% at taxable incomes of $10,000,000. The purpose of the 38% tax bracket is to wipe out the effect of the 34% tax bracket and to raise the average tax rate to 35%. This is accomplished at taxable income of $18,333,333. The income tax on $18,333,333 of taxable income is $6,416,667, which results in an average tax rate of 35% ($6,416,667/$18,333,333). Thereafter, the tax rate is reduced back to 35%.

15.2.1 Interest and Dividend Income

Interest income is taxed as ordinary income at the regular corporate tax rate. Corporate income is subject to *double taxation.* A corporation pays income tax on its taxable income, and when the corporation pays dividends to its individual shareholders, the dividends are subject to a second tax.

If a corporation owns stock in another corporation, then the income of the *subsidiary* corporation could be subject to triple taxation (income tax paid by the subsidiary, parent, and the individual shareholder). To avoid this result, corporate shareholders are entitled to reduce their income by a portion of the dividends received in a given year. Generally, the amount of the reduction depends upon the percentage of the stock of the subsidiary corporation owned by the *parent* corporation as shown below:

Percentage of Ownership by Corporate Shareholder	Deduction Percentage
Less than 20%	70%
20% or more, but less than 80%	80%
80% or more	100%

Example 15.5 — ABC Corporation owns 2% of the outstanding stock of XYZ Corporation, and ABC Corporation receives dividends of $10,000 in a given year from XYZ Corporation. As a result of these dividends, ABC Corporation will have ordinary income of $10,000 and an offsetting dividends received deduction of $7,000 (70% × $10,000), which results in a net $3,000 being subject to federal income tax. If ABC Corporation is in the 35% tax marginal tax bracket, its tax liability on the dividends is $1,050 (35% × $3,000). As a result of the dividends received deduction, these dividends are taxed at an effective federal tax rate of 10.5%.

15.2.2 Interest and Dividends Paid

Interest paid is a tax-deductible business expense. Thus, interest is paid with *before-tax* dollars. Dividends on stock (common and preferred), however, are not deductible and are therefore paid with *after-tax* dollars. This means that our tax system favors debt financing over equity financing.

Example 15.6 — Yukon Company has an operating income of $125,000, pays interest charges of $50,000, and pays dividends of $40,000. The company's taxable income is:

$125,000	(operating income)
–50,000	(interest charge, which is tax-deductible)
$75,000	(taxable income)

The tax liability, as calculated in Example 15.2, is $13,750 ($7,500 + $6,250). Note that dividends are paid with after-tax dollars.

15.2.3 Operating Loss Carryback and Carryforward

If a company has an operating loss, the loss may be applied against income in other years. The loss can be carried back 3 years and then forward 15 years. The corporate taxpayer may elect to first apply the loss against the taxable income in the 3 prior years. If the loss is not completely absorbed by the profits in these 3 years, it may be carried forward to each of the 15 following years. At that time, any loss remaining may no longer be used as a tax deduction.

To illustrate, 1999 operating loss may be used to recover, in whole or in part, the taxes paid during 1996, 1997, and 1998. If any part of the loss remains, this amount may be used to reduce taxable income, if any, during the 15-year period of 2000 through 2014. The corporation may choose to forego the loss carryback, and to instead carry the net operating loss to future years only.

Example 15.7 — The Loyla Company's taxable income and associated tax payments for the years 1994 though 2001 are presented below:

Year	Taxable Income ($)	Tax Payments ($)
1996	100,000	22,250
1997	100,000	22,250
1998	100,000	22,250
1999	(700,000)	0
2000	100,000	22,250
2001	100,000	22,250
2002	100,000	22,250
2003	100,000	22,250

In 1999, Loyla Company had an operating loss of $700,000. By carrying the loss back 3 years and then forward, the firm was able to "zero-out" its before-tax income as follows:

Year	Income Reduction ($)	Remaining 1999 Net Operating Loss ($)	Tax Savings ($)
1996	$100,000	$600,000	$ 22,250
1997	100,000	500,000	22,250
1998	100,000	400,000	22,250
1999	0	400,000	0
2000	100,000	300,000	22,250
2001	100,000	200,000	22,250
2002	100,000	100,000	22,250
2003	100,000	0	22,250
Total	$700,000		$157,500

As soon as the company recognized the loss of $700,000 in 1999, it was able to file for a tax refund of $66,750 ($22,250 + $22,250 + $22,500) for the years 1996 through 1998. It then carried forward the portion of the loss not used to offset past income and applied it against income for the next four years, 2000 through 2003.

15.2.4 Capital Gains and Losses

Capital gains and losses are a major form of corporate income and loss. They may result when a corporation sells investments and/or business property (not inventory). If depreciation has been taken on the asset sold, then part or all of the gain from the sale may be taxed as ordinary income.

Like all taxpayers, corporations net any capital gains and capital losses that they have. Corporations include any net capital gains as part of their taxable income. The maximum tax rate on the capital gains of individuals is 28%. Unlike individuals, corporations pay tax on their capital gains at the same rate as any other income. If an individual has a net capital loss, the individual may deduct up to $3,000 of that loss in the year incurred. The remaining capital loss is carried forward to future years indefinitely. Unlike individuals, corporations may not deduct any net capital losses. Instead, corporations may carry back the net capital loss to the three previous years and/or carry forward the net capital to the next five years. These capital loss carrybacks and carryforwards may be used to offset net capital gains in the past and/or future years.

15.2.5 Modified Accelerated Cost Recovery System (MACRS)

For all assets acquired after 1986, depreciation for tax purposes (cost recovery) is calculated using the Modified Accelerated Cost Recovery System (MACRS). MACRS was discussed in depth in Chapter 12.

15.2.6 Alternative "Pass Through" Tax Entities

As noted above, a disadvantage of corporations, compared to other forms of doing business (e.g., general partnerships), is double taxation. The net income of a corporation is taxed to the corporation. Later, should the corporation distribute that income to its shareholders, the distribution is taxed a second time to the recipient shareholders. Despite this disadvantage, corporations are popular because they have many advantages, including the fact that the liability of their shareholders (who are active in their business) for corporate debts is generally limited to the shareholders' investment in the corporations.

Two entities have developed (S corporations and limited liability companies), which allow investors to have limited liability and yet avoid double taxation. With these entities, owners of the entities are taxed on their share of the entities' income. Later, when that income is distributed to the owners, the distribution can be tax-free.

The importance of avoiding double taxation can be seen in the following example. Assume that a business has $100,000 of net income, and it has one shareholder, which is in the 28% marginal tax bracket. Assume that the business is either a corporation or a pass-through entity:

	Corporation	Pass-Through Entity
Entity's taxable income:	$100,000	$100,000
Tax on entity level:	(22,250)	(0)
Distribution to owner:	77,750	100,000
Tax on owner:	(21,770)	(28,000)
After-tax distribution:	$ 55,980	$ 72,000

Double taxation costs the investor $16,020 or approximately 16% in the above example. This percentage increases as the corporation's marginal tax rate increases.

Generally, the pass-through entity merely files an informational tax return with the Internal Revenue Service (IRS), and informs its owners of their share of the entity's taxable income or loss. The owners will be taxed on their share of the corporation's income. Afterwards, the distribution of any accrued income to the owners generally is tax-free.

15.2.6.1 S Corporations

Corporations, if they meet certain requirements, may elect to be taxed as S corporations. This is merely a tax classification, and corporations that make this election are still treated as general corporations for other legal purposes. In order to qualify for this treatment, the corporation must make a timely election with the IRS. In addition, the corporation must meet other requirements which were discussed in Chapter 1.

If an S corporation fails to meet any of the specified requirements, or if it voluntarily chooses to do so, its S corporation status will terminate. Upon termination, the corporation will be taxed as a general corporation. Generally, after such a termination, the corporation must wait five years before it may elect S corporation tax treatment again.

15.2.6.2 Limited Liability Companies

Limited liability companies (LLCs) are a relatively recent development. As was discussed in Chapter 1, LLCs are typically not permitted to carry on certain service businesses (e.g., law, medicine, and accounting). According to recent IRS regulations, an LLC may elect whether it wishes to be taxed as a corporation or as a pass-through entity (a partnership). Provided that the appropriate election is made, then the LLC will enjoy pass-through status.

15.3 FOREIGN TAX CREDIT

A credit is permitted for income taxes paid to a foreign country. However, the foreign tax credit cannot be used to reduce the U.S. tax liability on income from U.S. sources. The allowable credit is calculated as follows:

$$\text{Foreign tax credit} = \frac{\text{Foreign source income}}{\text{Worldwide income}} \times \text{U.S. liability}$$

Example 15.8 — In 2002 your company had worldwide taxable income of \$675,000 and tentative U.S. income tax of \$270,000. The company's taxable income from business operations in Country X was \$300,000, and foreign income taxes charged were \$135,000 stated in U.S. dollars. The credit for foreign income taxes that can be claimed on the U.S. tax return for 2001 is as follows:

$$\text{Foreign tax credit} = \frac{\$300{,}000}{\$675{,}000} \times \$270{,}000 = \$120{,}000$$

15.4 CONCLUSION

In performing your departmental responsibilities, it is important to understand the tax structure and its implications for your company. You must be familiar with the tax deductions, tax strategies, and depreciation if you are to obtain the most tax benefit for the company. You should also be aware of the tax implications regarding S corporations, capital gains and losses, and foreign tax credits. However, you should seek professional tax advice before committing funds to new projects and making specific financial and investment decisions.

Part IV

Obtaining Funds

16 What to Know About Short-Term Financing

"Short-term financing" refers to financing that will be repaid in 1 year or less. It may be used to meet seasonal and temporary fluctuations in funds position or to meet your permanent needs. For instance, short-term financing may be used to provide additional working capital, to finance current assets (e.g., receivables and inventory), or to provide interim financing for a long-term project (e.g., acquisition of a plant and equipment).

When compared to long-term financing, short-term financing has several advantages: it is easier to arrange, less expensive, and more flexible. The drawbacks of short-term financing are that interest rates fluctuate more often, refinancing is frequently required, there is a risk of not being able to repay, and delinquent repayment may be detrimental to your credit rating.

Sources of short-term financing include trade credit, bank loans, bankers' acceptances, finance company loans, commercial paper, receivable financing, and inventory financing. A particular source may be more appropriate in a given circumstance, and some are more desirable than others because of interest rates or collateral requirements.

You should consider the merits of the different alternative sources of short-term financing. The factors bearing upon the selection of a particular source include:

1. Cost.
2. Effect on credit rating. Some sources of short-term financing may negatively impact your credit rating (e.g., factoring accounts receivable).
3. Risk. Consider the reliability of the source of funds for future borrowing. If you are materially affected by outside forces, you need more stability and reliability in financing.
4. Restrictions. Certain lenders may impose restrictions, such as requiring a minimum level of working capital.
5. Flexibility. Certain lenders are more willing than others to work with you (e.g., to periodically adjust the amount of funds needed).
6. Expected money market conditions (e.g., future interest rates).
7. Inflation rate.
8. Profitability and liquidity positions.
9. Stability and maturity of your operations.
10. Tax rate.

16.1 HOW TO USE TRADE CREDIT

Trade credit (accounts payable) refers to balances owed to suppliers. It is a spontaneous (recurring) financing source, since it comes from normal operations, and it is the least expensive form of financing inventory. Trade credit has many advantages: it is readily available, since suppliers want business; collateral is not required; interest is typically not demanded or, if so, the rate is minimal; it is convenient; and trade creditors are frequently lenient if you get into financial trouble. If you have liquidity difficulties, you may be able to stretch (extend) accounts payable; however, this may lower your credit rating and may eliminate any cash discount offered.

Example 16.1 — You purchase $500 worth of merchandise per day from suppliers. The terms of purchase are net/60, and you pay on time. How much is your accounts payable balance?

$$\$500 \text{ per day} \times 60 \text{ days} = \$30{,}000$$

16.2 CASH DISCOUNTS

You should take advantage of a cash discount offered on the early payment of accounts payable because the failure to do so may result in a high opportunity cost. The cost of not taking a discount equals:

Discount lost	×	**360**	=	**Opportunity Cost (%)**
Dollar proceeds you have use of by not taking the discount		Number of days you have use of the money by not taking the discount		

Example 16.2 — You buy $1,000 in merchandise on terms of 2/10, net/30. You fail to take the discount, paying the bill on day 30. The opportunity cost is:

$$\frac{\$20}{\$980} \times \frac{360}{20} = 36.7\%$$

You should take the discount, even if you needed to borrow the money from the bank, since the interest rate on a bank loan would be far less than 36.7%.

16.3 WHEN ARE BANK LOANS ADVISABLE?

Most banking activities are conducted by commercial banks, although all institutions (e.g., savings and loan associations, credit unions) provide banking services as well. Commercial banks allow you to operate with minimal cash and still be confident of planning activities even in times of uncertainty. These banks favor short-term loans; however, loans in excess of 1 year may be given (see Chapter 17).

To be eligible for a bank loan, you must have sufficient equity and good liquidity. A banker gathers information regarding your business operations and often expresses the bank's opinions of your practices. Depending on your financial standing, the bank may take a "soft" or "hard" position including warnings of not renewing loans or demanding immediate payment if you do not agree to its suggestions. If your company is large, a group of banks may form a consortium to furnish the desired level of capital. One bank is appointed the prime negotiator.

When a short-term bank loan is taken, you usually issue a borrower's statement (see Exhibit 16.1) and sign a note. This note is a written statement that you agree to repay the loan at the due date. A note payable may be paid at maturity or in installments and consists of principal and interest.

Exhibit 16.1 — Typical Statement Required of Corporate Borrower

To ABC Bank:

To obtain credit from time to time with you for our negotiable paper, we provide financial statements as of December 31, 2001. We will notify you immediately of any significant unfavorable change in our financial position, and, in the absence of such notice or of a new and full written statement, this may be construed as a continuing statement and substantially accurate; and it is hereby agreed that upon requesting further credit, this statement shall have the same force as if delivered as an original statement of our financial position at the time when further credit is needed.

If any judgment is entered or any legal action is commenced against the undersigned, or if the undersigned becomes financially troubled, or on the failure of the undersigned to notify you of any important change in the financial status of the undersigned, or if the undersigned assigns any accounts or transfers any assets which you deem materially affect the business, then all obligations held by you upon which the undersigned is responsible shall at your option, be immediately due, notwithstanding the date of payment as fixed by the obligation then held by you, and any credit balance of the undersigned may be applied by you in satisfaction of any such debt.

A bank loan is not a source of spontaneous financing as is trade credit. You must apply for loans, and lenders must grant them. Without additional funds, you may have to restrict your plans; therefore, as your need for funds change, you should alter your borrowings from banks. For example, a self-liquidating (seasonal) loan may be used to pay for a temporary increase in accounts receivable or inventory. As soon as the assets realize cash, you repay the loan.

Loans, of course, earn interest, and the prime interest rate is the lowest interest rate applied to short-term loans from a bank. Banks charge only their most credit-worthy clients the prime rate; other borrowers are charged higher interest rates. You must realize that the prime interest rate is a negotiating point for borrowed funds. Your interest rate will probably be higher depending upon your risk.

Bank financing may take any of the following forms:

- Unsecured loans
- Secured loans
- Lines of credit
- Installment loans

16.3.1 Are You Eligible for an Unsecured Loan?

Most short-term, unsecured (no collateral) loans are self-liquidating. This kind of loan is recommended if you have an excellent credit rating. It is appropriate if you have immediate cash and can either repay the loan in the near future or can quickly obtain longer-term financing. Seasonal cash shortfalls and desired inventory buildups are reasons to use an unsecured loan. There are two disadvantages to this kind of loan: (1) it carries a higher interest rate than a secured loan and (2) payment in a lump sum is required.

16.3.2 What Will You Give to Obtain a Secured Loan?

If your credit rating is deficient, the bank may lend money only on a secured basis — that is, with some form of collateral behind the loan. Collateral may take many forms including inventory, marketable securities, or fixed assets. In some cases, even though you are able to obtain an unsecured loan, you may still give collateral to get a lower interest rate.

16.3.3 What Line of Credit Can You Get?

Under a line of credit, the bank agrees to lend money on a recurring basis up to a specified amount (see Exhibit 16.2). Credit lines are typically established for a 1-year period and may be renewed annually.

Exhibit 16.2 — Sample Letter Extending Line of Credit

Grace Bank
New York, N.Y.

March 16, 2001

Mr. Jim Balk
Treasurer
Harris Manufacturing Corporation
New York, NY

Dear Mr. Balk,

Based upon our review of your year-end audited financial statements, we renew your $2 million unsecured line of credit for next year. Borrowings under this line will be at a rate of 1% over the prime rate.

This line is subject to your company maintaining its financial position and being out of bank debt for at least 90 days during the fiscal year.

Sincerely,
Bob Jones
Vice President

The advantages of a line of credit are the easy and immediate access to funds during tight money market conditions and the ability to borrow only as much as needed and repay immediately when cash is available. The disadvantages relate to the collateral requirements and the additional financial information that must be presented to the bank. Also, the bank may place restrictions upon you, such as a ceiling on capital expenditures or the maintenance of a minimum level of working capital. Further, the bank typically charges a commitment fee on the amount of the unused credit line.

When you borrow under a line of credit, you may be required to maintain a deposit with the bank that does not earn interest. This deposit is referred to as a *compensating balance* and is stated as a percentage of the loan. The compensating balance effectively increases the cost of the loan. A compensating balance may also be placed on the unused portion of a line of credit, in which case the interest rate would be reduced.

Example 16.3 — You may borrow $200,000 and are required to keep a 12% compensating balance. You also have an unused line of credit in the amount of $100,000, for which a 10% compensating balance is required. The minimum balance that must be maintained is:

$$(\$200{,}000 \times .12) + (\$100{,}000 \times .10) = \$24{,}000 + \$10{,}000$$
$$= \$34{,}000$$

A line of credit is typically decided upon prior to the actual borrowing. In the days between the arrangement for the loan and the actual borrowing, interest rates will change. Therefore, your agreement with the bank will stipulate it is at the prime interest rate prevailing when the loan is extended plus a risk premium. Note that the prime interest rate is not known until you actually borrow the money. For example, if you arrange for a line of credit when the prime interest rate is 8% and the interest rate on the loan is stipulated at 2% above prime, but the prime interest rate is 10% on the day you actually borrow the money, you will be charged an interest rate of 12%.

Be on guard! Some banks may test your financial capability by requiring you to "clean up" the loan — that is, repay it for a brief time during the year (e.g., for 1 month). The payment shows the bank that the loan is actually seasonal rather than permanent. If you are unable to repay a short-term loan, you should probably finance with long-term funds.

16.3.4 What Is an Installment Loan?

An installment loan requires monthly payments. When the principal on the loan decreases sufficiently, refinancing can take place at lower interest rates. The advantage of this kind of loan is that it may be tailored to satisfy seasonal financing needs. The disadvantage is that you may have a cash problem in a particular month but still have to pay the monthly installment payment.

16.3.5 How Do You Compute Interest?

Interest on a loan may be paid either at maturity (ordinary interest) or in advance (discounting the loan). When interest is paid in advance, the loan proceeds are reduced and the effective (true) interest cost is increased.

Example 16.4 — You borrow $30,000 at 16% interest per annum and repay the loan 1 year later. The interest is:

$$\$300{,}000 \times .16 = \$4{,}800$$

The effective interest rate is 16% ($4,800/$30,000).

Example 16.5 — Assume the same facts as in Example 16.4, except the note is discounted. The proceeds of this loan are smaller than in the previous example.

$$\begin{aligned} \text{Proceeds} &= \text{Principal} - \text{Interest} \\ &= \$30{,}000 - \$4{,}800 \\ &= \$25{,}200 \end{aligned}$$

The true interest rate for this discounted loan is:

$$\text{Effective interest rate} = \frac{\text{Interest}}{\text{Proceeds}} = \frac{\$4{,}800}{\$25{,}000} = 19\%$$

Example 16.6 — Bank A will give you a 1-year loan at an interest rate of 20% payable at maturity, while Bank B will lend on a discount basis at a 19% interest rate. The effective rate for Bank B is calculated as follows:

$$\frac{19\%}{81\%} = 23.5\%$$

Bank A's 20% rate is the lowest effective rate.

When a loan has a compensating balance requirement, the proceeds received are decreased by the amount of the compensating balance. The compensating balance will increase the effective interest rate.

Example 16.7 — The effective interest rate for a 1-year, $600,000 loan that has a nominal interest rate of 19%, with interest due at maturity and with a 15% compensating balance required, is calculated as follows:

$$\begin{aligned} \frac{\text{Interest rate} \times \text{Principal}}{\text{Percentage of proceeds available} \times \text{Principal}} &= \frac{.19 \times \$600{,}000}{(1.00 - .15) \times \$600{,}000} \\ &= \frac{\$114{,}000}{\$510{,}000} \\ &= 22.4\% \end{aligned}$$

Example 16.8 — Assume the same facts as in Example 16.7, except that the loan is discounted. The effective interest rate is:

$$\frac{\text{Interest rate} \times \text{Principal}}{\text{Percentage of proceeds available} \times \text{Principal}} = \frac{0.19 \times \$600{,}000}{(0.85 \times \$600{,}000) - \$114{,}000}$$

$$= \frac{\$114{,}000}{\$396{,}000}$$

$$= 28.8\%$$

Example 16.9 — You have a credit line of $400,000, but you must maintain a compensating balance of 13% on outstanding loans and a compensating balance of 10% on the unused credit. The interest rate on the loan is 18%. You borrow $275,000. The required compensating balance is:

.13 × $275,000	$35,750
.10 × 125,000	12,500
	$48,250

The effective interest rate (with line of credit) is:

$$\frac{\text{Interest rate (on loan)} \times \text{Principal}}{\text{Principal} - \text{Compensating balance}} = \frac{0.18 \times \$275{,}000}{\$275{,}000 - \$48{,}250}$$

$$= \frac{\$49{,}500}{\$226{,}750}$$

$$= 21.8\%$$

On an installment load, the effective interest rate computation is more complicated. Assuming a 1-year loan payable in equal monthly installments, the effective rate is:

$$\frac{\text{Interest}}{\text{Average loan balance}}$$

The average loan balance is one half the loan amount. The interest is computed on the face amount of the loan.

Example 16.10 — You borrow $40,000 at an interest rate of 10% to be paid in 12 monthly installments. The average loan balance is:

$$\frac{\$40{,}000}{2} = \$20{,}000$$

The effective interest rate is:

$$\frac{\$4{,}000}{\$20{,}000} = 20\%$$

Example 16.11 — Assume the same facts as in Example 16.10, except that the loan is discounted. The interest of $4,000 is deducted in advance, so the proceeds received are:

$$\$40,000 - \$4,000 = \$36,000$$

The average loan is:

$$\frac{\$36,000}{2} = \$18,000$$

The effective interest rate is:

$$\frac{\$4,000}{\$18,000} = 22.2\%$$

16.4 WHAT SHOULD YOU KNOW WHEN DEALING WITH A BANKER?

When applying for a loan, you should know that the loan officer will prepare financial ratios of your liquidity (ability to meet short-term debt) and will ask questions to determine your managerial competency. Since the banker will want to know whether you will be able to repay the loan, he or she will want to see your cash flow forecasts. You will make a good impression if you have the financial statements and documents readily available and are prepared to discuss your financial condition, particularly being able to explain ratios that depart from industry norms.

Banks are anxious to lend money to meet self-liquidating, cyclical business needs. A short-term bank loan is an inexpensive way to obtain funds to satisfy working capital requirements during the business cycle. However, you must be able to explain what the needs are in an intelligent manner so the banker has confidence in your abilities.

In a long-term banking relationship, bankers offer professional assistance in addition to financing. They may recommend ways to improve the quality of business policies, to enter new markets, etc.

16.5 WHAT ARE BANKER'S ACCEPTANCES?

A *banker's acceptance* is a draft, drawn by you and accepted by a bank, that orders payment to a third party at a later date. The creditworthiness of the draft is of good quality because it has the backing of the bank, not the client. It is, in essence, a debt instrument created out of a self-liquidating transaction. Banker's acceptances are often used to finance the shipment handling of both domestic and foreign merchandise. Acceptances act as short-term financing because they typically have maturities of 180 days.

Example 16.12 — A U.S. oil refiner arranges with its U.S. commercial bank for a letter of credit to a Saudi Arabian exporter with whom the U.S has undertaken a transaction. The letter of credit regarding the shipment states that the exporter can draw a time draft for a given amount on the U.S. bank. The exporter draws a draft on the bank and negotiates it with a local Saudi Arabian bank, re: payment. The Saudi Arabian bank sends the draft to the U.S., the latter accepts the draft, and there is an acceptance to meet the maturity date.

16.6 ARE YOU FORCED TO TAKE OUT A COMMERCIAL FINANCE COMPANY LOAN?

When credit is unavailable from a bank, you may have to go to a commercial finance company (e.g., CIT Financial). The finance company loan has a higher interest rate than a bank, and generally it is secured. Typically, the amount of collateral placed will be greater than the balance of the loan. Collateral includes accounts receivable, inventories, and fixed assets. Commercial finance companies also finance the installment purchases of industrial equipment by firms. A portion of their financing is sometimes obtained through commercial bank borrowing at wholesale rates.

16.7 ARE YOU FINANCIALLY STRONG ENOUGH TO BE ABLE TO ISSUE COMMERCIAL PAPER?

Commercial paper can be issued only if you possess a very high credit rating. Therefore, the interest rate is less than that of a bank loan, typically 0.5% below the prime interest rate. Commercial paper is unsecured and sold at a discount (below face value). The maturity date is usually less than 270 days; otherwise, registration with the Securities and Exchange Commission (SEC) is needed. When a note is sold at a discount, it means that interest is immediately deducted from the face of the note by the creditor, but you will pay the full face value. Commercial paper may be issued through a dealer or directly placed to an institutional investor.

The benefits of commercial paper are that no security is required, the interest rate is typically less than through bank or finance company borrowing, and the commercial paper dealer often offers financial advice. The drawbacks are that commercial paper can be issued only by large, financially sound companies, and commercial paper dealings relative to bank dealings are impersonal.

Example 16.13 — You issue $500,000 of commercial paper every 2 months at a 13% rate. There is a $1,000 placement cost each time. The percentage cost of the commercial paper is:

Interest ($500,000 × .13)	$65,000
Placement cost ($1,000 × 6)	6,000
Cost	$71,000

$$\text{Percentage cost of commercial paper} = \frac{\$71{,}000}{\$500{,}000} = 14.2\%$$

16.8 SHOULD RECEIVABLES BE USED FOR FINANCING?

In accounts receivable financing, the accounts receivable are the security for the loan as well as the source of repayment. Accounts receivable financing generally takes place under the following conditions:

1. Receivables are a minimum of $25,000.
2. Sales are a minimum of $250,000.
3. Individual receivables are at a minimum of $100.
4. Receivables apply to selling merchandise rather than rendering services.
5. Customers are financially strong.
6. Sales returns are not great.
7. Title to the goods passes to the buyer at shipment.

Receivable financing has several advantages, including avoiding the need for long-term financing and obtaining a recurring cash flow. Accounts receivable financing has the drawback of high administrative costs when there are many small accounts; however, with the use of personal computers these costs can be curtailed.

Accounts receivable may be financed under either a factoring or assignment (pledging) arrangement. *Factoring* is the outright sale of accounts receivable to a bank or finance company *without recourse*. The purchaser of the factoring arrangement (also called the *factor*) takes all credit and collection risks. The proceeds received are equal to the face value of the receivables less the commission charge, which is usually 2% to 4% higher than the prime interest rate. The cost of the factoring arrangement is the factor's commission for credit investigation, interest on the unpaid balance of advanced funds, and a discount from the face value of high-credit-risk receivables. Remissions by customers are made directly to the factor.

The advantages of factoring are immediate availability of cash, reduction in overhead because the credit examination function is no longer needed, obtaining of financial advice, receipt of advances as required on a seasonal basis, and strengthening of the balance sheet position.

The disadvantages of factoring include both the high cost and the negative impression left with customers due to the change in ownership of the receivables. Also, factors may antagonize customers by their demanding methods of collecting delinquent accounts.

In an *assignment* (pledging), there is no transfer of the ownership of the accounts receivable. Receivables are given to a finance company *recourse*. The finance company usually advances between 50% and the face value of the receivables in cash. You are responsible for charges, interest on the advance, and any resulting bad debt losses. Remissions continue to be made directly to you.

The assignment of accounts receivable has the advantages of availability of cash, cash advance accessibility on a seasonal basis, and avoidance of negative customer feelings. The disadvantages include the high cost, the continuance of the clerical function associated with accounts receivable, and the bearing of all credit risk.

You must also be aware of the impact of a change in accounts receivable policy on the cost of financing receivables. When accounts receivable are financed, the cost of financing may rise or fall under different circumstances. For example, when credit standards are relaxed, costs rise; when defaults are given to the finance company, costs decrease; and when the minimum invoice amount of a credit sale is increased, costs decline. You should compute the costs of accounts receivable financing and select the least expensive alternative.

Example 16.14 — You have $120,000 per month in accounts receivable that a factor will purchase from you, advancing you up to 80% of the receivables for an annual charge of 14% and a 1.5% fee on receivables purchased. The cost of this factoring arrangement is:

Factor fee [.015 × ($120,000 × 12)]	$21,600
Cost of borrowing [.014 × ($120,000 × .8)]	13,440
Total cost	$35,040

Example 16.15 — Your factor charges a 3% fee per month. The factor lends you up to 75% of receivables purchased for an additional 1% per month. Your credit sales are $400,000 per month. As a result of the factoring arrangement, you save $6,500 per month in credit costs and a bad debt expense of 2% of credit sales.

XYZ Bank has offered an arrangement to lend you up to 75% of the receivables. The bank will charge 2% per month interest plus a 4% processing charge on receivable lending.

The collection period is 30 days. If you borrow the maximum per month, should you stay with the factor or switch to XYZ Bank? The cost of the factor is:

Purchased receivables (.03 × $400,000)	$12,000
Lending fee [.01 × (.75 × $400,000)]	3,000
Total cost	$15,000

The cost of the bank financing is:

Interest [.02 × (.75 × $400,000)]	$ 6,000
Processing charge [.04 × (.75 × $400,000)]	12,000
Additional cost of not using the factor	
Credit costs	6,500
Bad debts (.02 × $400,000)	8,000
Total cost	$32,500

You should stay with the factor.

16.9 SHOULD INVENTORIES BE USED FOR FINANCING?

Inventory financing typically occurs when you have completely used your borrowing capacity on receivables. The inventory itself must be good security for a loan, it must consist of marketable, nonperishable, and standardized goods that have quick turnover, and it should not be subject to rapid obsolescence. Inventory financing should consider the price stability of the merchandise and the costs of selling it.

The advance is high when there is marketable inventory. In general, the financing of raw materials and finished goods is about 75% of their value. The interest rate is approximately 3 to 5% over the prime interest rate. This high interest rate is one of the disadvantages of inventory financing. The restrictions placed on inventory represent another drawback.

The types of inventory financing include a floating (blanket) lien, warehouse receipt, and trust receipt. With a *floating lien*, the creditor's security lies in the aggregate inventory rather than in its components. Even though you sell and restock, the lender's security interest continues. With a *warehouse receipt*, the lender receives an interest in the inventory stored at a public warehouse; however, the fixed costs of this arrangement are high. There may be a field warehouse arrangement in which the warehouse sets up a secured area directly at your location. You have access to the goods but must continually account for them. With a *trust receipt* loan, the creditor has title to the goods but releases them to you to sell on the creditor's behalf. As goods are sold, you remit the funds to the lender. A good example of trust receipt use is in automobile dealer financing. The drawback of the trust receipt arrangement is that a trust receipt must be given for specific items.

A *collateral certificate* may be issued by a third party to the lender guaranteeing the existence of pledged inventory. The advantage of a collateral certificate is flexibility because merchandise does not have to be segregated or possessed by the lender.

Example 16.16 — You wish to finance $500,000 of inventory. Funds are required for 3 months. A warehouse receipt loan may be taken at 16% with a 96% advance against the inventory's value. The warehousing cost is $4,000 for the 3-month period. The cost of financing the inventory is:

Interest [.16 × .90 × $500,000 × (3/12)]	$18,000
Warehousing cost	4,000
Total cost	$22,000

Example 16.17 — You show growth in operations but are presently experiencing liquidity difficulties. Six large financially sound companies are responsible for 75% of your sales. Sales amount to $1,800, and net income is $130,000. On the basis of the financial information that is provided in the following balance sheet, would you be on receivables or inventory?

Receivable financing can be expected since a high percent is made to only six large financially strong companies. Receivables will thus show good collectibility. It is also easier to control accounts.

Inventory financing is not likely, due to the high p completed items. Lenders are reluctant to finance inventory when a large work-in-progress balance exists because the goods will be difficult to further process and sell by lenders.

Assets

Current assets		
Cash	$27,000	
Receivables	380,000	
Inventory (consisting of 55% of work-in-progress)	320,000	
Total current assets		$727,000
Fixed assets		250,000
Total assets		$977,000

Liabilities and Stockholders' Equity

Current liabilities	$260,000	
Accounts payable	200,000	
Loans payable	35,000	
Total current liabilities		$495,000
Non-current liabilities		
Bond payable		110,000
Total liabilities		$605,000
Stockholders' equity		
Common stock	$250,000	
Retained earnings	122,000	
Total stockholders' equity		372,000
Total liabilities and stockholders' equity		$977,000

16.10 WHAT OTHER ASSETS MAY BE USED FOR FINANCING?

Inventory and receivables are not the only assets that may be used as security for short-term bank loans. Real estate, plant and equipment, cash surrender value of life insurance policies, and securities may also be used. Also, lenders are typically willing to advance a high percentage of the market value of bonds, and loans may also be made based on a guaranty of a third party.

16.11 CONCLUSION

In short-term financing, the best financing tool should be used to meet your objectives. The financing instrument depends upon your particular circumstances. Consideration is given to such factors as cost, risk, reactions, stability of operations, and tax rate. Sources of short-term financing include trade credit, bank loans, banker's acceptances, finance company loans, commercial paper, receivable financing, and inventory financing. If you think you will be short of cash during certain times, arrange for financing (e.g., a line of credit) in advance instead of waiting for an emergency.

Table 16.1 presents a summary of the major features of short-term financing sources, and Figure 16.1 illustrates temporary and permanent financing.

TABLE 16.1
Summary of Short-Term Financing Sources

Types of Financing	Source	Cost or Terms	Features
Spontaneous Sources			
Accounts payable	Suppliers	No explicit cost but there is an opportunity cost if a cash discount for early payment is not taken. Companies should take advantage of the discount offered.	The main source of short-term financing typically on terms of 0 to 120 days.
Accrued expenses	Employees and tax agencies	None	Expenses incurred but not yet paid (e.g., accrued wages payable, accrued taxes payable).
Unsecured Sources			
Bank loans			
1. Single payment note	Commercial banks	Prime interest rate plus risk premium. The interest rate may be fixed or variable. Unsecured loans are less costly than secured loans.	A single-payment loan to satisfy a funds shortage to last a short time period.
2. Lines of credit	Commercial banks	Prime interest rate plus risk premium. The interest rate may be fixed or variable. A compensating balance is typically required. The line of credit must be "cleaned up" periodically.	An agreed upon borrowing limit for funds to satisfy seasonal needs.
Commercial paper	Commercial banks, insurance companies, other financial institutions, and other companies	A little less than the prime interest rate.	Unsecured short-term note of financially strong companies.
Secured Sources			
Accounts receivable as collateral			
1. Pledging	Commercial banks and finance companies	Typically 2 to 5% above prime plus fees (usually 2 to 3%). Low administrative costs. Advances typically ranging from 60 to 85%.	Qualified accounts receivable accounts serve as collateral upon collection of the account; the borrower remits to the lender. Customers are not notified of the arrangement. With recourse means that the risk of nonpayment continues to be borne by the company.

TABLE 16.1 (continued)
Summary of Short-Term Financing Sources

Types of Financing	Source	Cost or Terms	Features
2. Factoring	Factors, commercial banks, and commercial finance companies	Typically a 2 to 3% discount from the face value of factored receivables. Interest on advances of almost 3% over prime. Interest on surplus balances held by factor of about ½% per month. Costs with factoring are higher than with pledging.	Certain accounts receivable are sold on a discount basis without recourse. This means that the factor bears the risk of nonpayment. Customers are notified of the arrangement. The factor provides more services than is the case with pledging.
Inventory collateral			
1. Floating liens	Commercial banks and commercial finance companies	About 4% above prime. Advance is about 40% of collateral value.	Collateral is all the inventory. There should be a stable inventory with many inexpensive items.
2. Trust receipts (floor planning)	Commercial banks and commercial finance companies	About 3% above prime. Advances ranging from 80 to 100% of collateral value.	Collateral is specific inventory that is typically expensive. Borrower retains collateral. Borrower remits proceeds to lender upon sale of the inventory.
3. Warehouse receipts	Commercial banks and commercial finance companies	About 4% above prime plus about a 2% warehouse fee. Advance of about 80% of collateral value.	Collateralized inventory is controlled by lender. A warehousing company issues a warehouse receipt held by the lender. The warehousing company acts as the lender's agent.

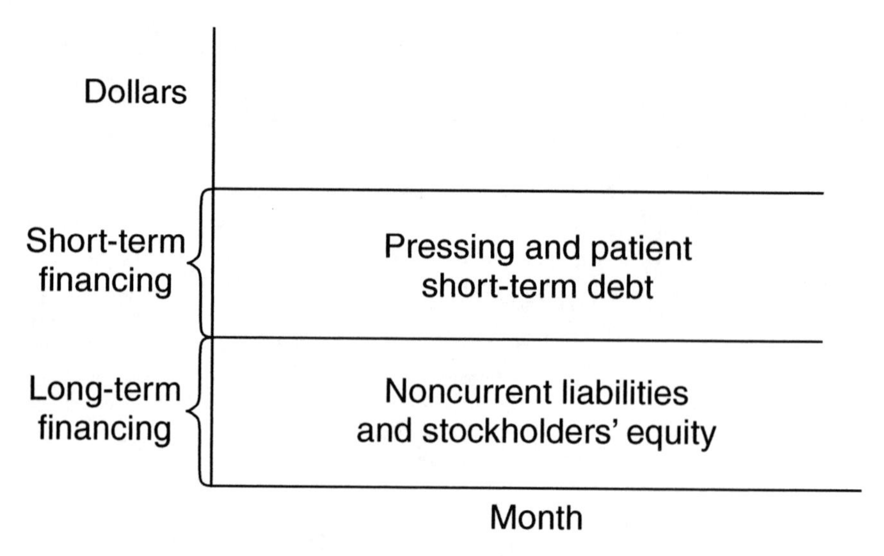

FIGURE 16.1 Illustrative financing.

17 Looking at Term Loans and Leasing

This chapter considers intermediate-term loans, primarily from banks and leasing arrangements to meet your financing needs. Intermediate-term loans include bank loans, revolving credit, insurance company term loans, and equipment financing.

17.1 INTERMEDIATE-TERM BANK LOANS

Intermediate-term loans are typically loans with a maturity of more than 1 year but less than 5 years. They are appropriate when short-term unsecured loans are not, such as when a business is acquired, new fixed assets are purchased, and long-term debt is required. If you wish to float long-term debt or issue common stock but conditions are unfavorable in the market, you may seek an intermediate loan to bridge the gap until long-term financing can be undertaken on favorable terms.

The interest rate on an intermediate-term loan is typically higher than that on a short-term loan because of the longer maturity period. The interest rate may be either fixed or variable (according to, for instance, changes in the prime interest rate). The cost of an intermediate-term loan changes with the amount of the loan and your financial strength.

Ordinary intermediate-term loans are payable in periodic, equal installments, except for the last payment, which may be higher (referred to as balloon payment). The schedule of loan payments should be based on the cash flow position to satisfy the debt. The periodic payment in a term loan is computed as follows:

$$\text{Periodic payment} = \frac{\text{Amount of loan}}{\text{Present value factor}}$$

Example 17.1 — You contract to repay a term loan in five equal year-end installments. The amount of the loan is \$150,000 and the interest rate is 10%. The payment each year is

$$\frac{\$150{,}000}{3.791} = \$39{,}567.40$$

where 3.791 = T_4 (10%, 5 years) = present value of annuity for 5 years at 10% (from Table 11.4).

Example 17.2 — You may want to determine the amount of loan, and what portion of the periodic payment represents principal and interest. You take out a term loan to be paid off in 20 year-end annual installments of $2,000 each. The interest rate is 12%. The amount of the loan is

$$\$2,000 = \frac{\text{Amount of Loan}}{7.469}$$

$$\text{Amount of loan} = \$2,000 \times 7.469 = \$14,938.00$$

where 7.469 = T_4(12%, 20 years) from Table 11.4.

The amortization schedule for the first 2 years is shown in the table:

Year	Payment	Interest[a]	Principal	Balance
0				$14,938.00
1	$2,000	$1,792.56	$207.44	$14,730.56
2	$2,000	$1,767.67	$232.33	$14,498.23

[a] 12% times the balance of the loan at the beginning of the year.

When taking out an intermediate-term loan, you will encounter several kinds of restrictive provisions that are designed to protect the lender. You may find that the agreement includes general provisions that are used in most agreements and vary depending upon your situation. Examples are working capital and cash dividend requirements. Or the agreement may contain routine (uniform) provisions that are employed in most agreements and are applied universally. Examples are the payment of taxes and the maintenance of proper insurance to ensure maximum lender protection. Lastly, you may encounter specific provisions that are tailored to a particular situation. Examples are the placing of limits on future loans and the carrying of adequate life insurance for executives.

The advantages of intermediate-term loans include the following:

1. Intermediate-term loans provide flexibility in that the term may be altered as your financing requirements change.
2. Financial information is kept confidential, since no public issuance is involved.
3. The loan may be arranged quickly, relative to a public offering.
4. These loans avoid the possible nonrenewal of a short-term loan.
5. Public flotation costs are not involved.

The disadvantages of intermediate-term loans are as follows:

1. Collateral and possible restrictive convenants are required, in contrast to commercial paper and unsecured short-term bank loans.
2. Budgets and financial statements may have to be submitted periodically to the lender.
3. "Kickers" or "sweeteners," such as stock warrants or a share of the profits, are sometimes requested by the bank.

17.2 USING REVOLVING CREDIT

Revolving credit, usually used for seasonal financing, may have a 3-year maturity, but the notes evidencing revolving credit are short, usually 90 days. The advantages of revolving credit are flexibility and ready availability. Within the time period of the revolving credit, you may renew a loan or enter into additional financing up to a maximum amount. There are typically fewer restrictions on revolving credit but at the cost of a slightly higher interest rate.

17.3 INSURANCE COMPANY TERM LOANS

Insurance companies and other institutional lenders may extend intermediate-term loans. Insurance companies typically accept loan maturity dates exceeding 5 years, but their rate of interest is often higher than that of bank loans. Insurance companies do not require compensating balances, but usually there is a payment penalty, which is typically not the case with a bank loan. You may take out an insurance company loan when you desire a longer maturity range.

17.4 FINANCING WITH EQUIPMENT

Equipment may serve as collateral for a loan. An advance is made against the market value of the equipment. The more marketable the equipment, the higher the advance will be. Also considered is the cost of selling the equipment. The repayment schedule is designed so that the market value of the equipment at any given time is in excess of the unpaid loan principal.

Equipment financing may be obtained from banks, finance companies, and manufacturers of equipment. Equipment loans may be secured by a chattel mortgage or a conditional sales contract. A *chattel mortgage* serves as a lien on property excluding real estate. In a *conditional sales contract*, the seller keeps title to the equipment until the buyer has satisfied the terms; otherwise, the seller will repossess the equipment. The buyer makes periodic payments to the seller over a specified time period. A conditional sales contract is generally used if you are a small company with a low credit rating.

17.5 LEASING

The parties in a lease are the *lessor* who legally owns the property, and the *lessee*, who uses it in exchange for making rental payments.

The following types of leases exist:

- **Operating (service) lease** — This type of lease includes both financing and maintenance services. You lease property owned by the lessor. The lessor may be the manufacturer of the asset or it may be a leasing company that buys assets from the manufacturer to lease to others. The lease payments under the contract are typically not adequate to recover the full cost of the property. Maintenance and service are provided by the lessor, and related costs are included in the lease payments. There usually exists a cancellation clause, which provides you with the right to cancel the contract and return the property prior to the expiration date of the agreement. The life of the contract is less than the economic life of the property.

- **Financial Lease** — This type of lease does not typically provide for maintenance services, is noncancelable, and the rental payments equal the full price of the leased property. The life of the contract approximates the life of the property.
- **Sale and leaseback** — With this lease arrangement, you sell an asset to another (usually a financial institution) and then lease it back. This allows you to obtain cash from the sale and still have the property for use.
- **Leveraged lease** — In a leveraged lease, a third party serves as the lender. Here, the lessor borrows money from the lender to buy the asset. The property is then leased to you.

Leasing has several advantages, including the following:

- Immediate cash outlay is not required.
- Typically, a purchase option exists, allowing you to obtain the property at a bargain price at the expiration of the lease. This provides the flexibility to make the purchase decision based on the value of the property at the termination date.
- The lessor's expert service is made available.
- Typically, fewer financing restrictions (e.g., limitations on dividends) are placed by the lessor than are imposed when obtaining a loan to purchase the asset.
- The obligation for future rental payment does not have to be reported on the balance sheet if the lease is considered an operating lease.
- Leasing allows you, in effect, to depreciate land, which is not allowed if land is purchased.
- In bankruptcy or reorganization, the maximum claim of lessors is 3 years of lease payments. With debt, creditors have a claim for the total amount of the unpaid financing.

There are several drawbacks to leasing, including the following:

- In the long run, the cost is higher than if the asset is bought.
- The interest cost of leasing is typically higher than the interest cost on debt.
- If the property reverts to the lessor at termination of the lease, you must either sign a new lease or buy the property at higher current prices. Also, the salvage value of the property is realized by the lessor.
- You may have to retain property no longer needed (i.e., obsolete equipment).
- You cannot make improvements to the leased property without the permission of the lessor.

You may want to determine the periodic payment required on a lease.

Example 17.3 — You enter into a lease for a $100,000 machine. You are to make 10 equal annual payments at year-end. The interest rate on the lease is 14%. The periodic payment equals

$$\frac{\$100{,}000}{5.2161} = \$19{,}171$$

where T_4(14%, 10 years) = the present value of an ordinary annuity factor for n = 10, i = 14% = 5.2161 (Table 11.4).

Example 17.4 — Use the same facts as those for Example 17.3, except that the annual payments are to be made at the beginning of each year. The periodic payment is computed as follows:

Year	Factor
0	1.0
1–9	4.946
	5.946

$$\frac{\$100{,}000}{5.946} = \$16{,}818$$

The interest rate associated with a lease agreement can also be computed. Divide the value of the leased property by the annual payment to obtain the factor, which is then used to find the interest rate with the help of an annuity table.

Example 17.5 — You leased $300,000 of property and are to make equal annual payments at year-end of $40,000 for 11 years. The interest rate associated with the lease agreement is

$$\frac{\$300{,}000}{\$40{,}000} = 7.5$$

Going to the present value of annuity table (T_4) and looking across the 11-years entry to a factor nearest to 7.5, we find 7.4987 at a 7% interest rate. Thus, the interest rate in the lease agreement is 7%.

The capitalized value of a lease can be found by dividing the annual lease payment by an appropriate present value of annuity factor.

Example 17.6 — Property is to be leased for 8 years at an annual rental payment of $140,000, payable at the beginning of each year. The capitalization rate is 12%. The capitalized value of the lease is

$$\frac{\text{Annual lease payment}}{\text{Present value factor}} = \frac{\$140{,}000}{1 + 4.5638} = \$25{,}163$$

17.6 LEASE-PURCHASE DECISION

Often you must decide whether it is better to buy an asset or to lease it. Present value analysis may be used to determine the cheapest alternative (see Chapter 12).

17.7 CONCLUSION

This chapter discussed intermediate-term loans, primarily from banks and leasing arrangements to meet your financing needs. Intermediate-term loans include bank loans, revolving credit, insurance company term loans, and equipment financing.

18 Deciding on Long-Term Financing

This chapter will provide you with a deeper understanding of the financing of your company. You will learn the fundamentals of equity and long-term debt financing. Equity financing consists of issuing preferred stock and common stock. Long-term debt financing consists primarily of issuing bonds. Long-term financing is often used to finance long-lived assets (e.g., land, plant) or construction projects.

First, the role of the investment banker is considered. Further, a comparison of public vs. private placement of securities is provided. The advantages and disadvantages of issuing long-term debt, preferred stock, and common stock are presented, as well. You will also learn what financing strategy is most appropriate, given a particular set of circumstances, for your company.

Figure 18.1 shows the general picture with financing assets through debt and equity.

18.1 INVESTMENT BANKING

Investment banking involves the sale of a security issue. Investment bankers conduct the following activities:

1. **Underwriting** — The investment banker buys a new security issue, pays the issuer, and markets the securities. The underwriter's compensation is the difference between the price at which the securities are sold to the public and the price paid to the issuing company.
2. **Distributing** — The investment banker markets your security issue.
3. **Advising** — The investment banker gives you advice regarding the optimal way to obtain funds. The investment banker is knowledgeable about the alternative sources of long-term funds, debt and equity markets, and Securities and Exchange Commission (SEC) regulations.
4. **Providing funds** — The investment banker provides funds to you during the distribution period.

When several investment bankers form a group because a particular issue is large and/or risky, they are termed a *syndicate*. A syndicate is a temporary association of investment bankers brought together for the purpose of selling new securities. One investment banker among the group will be selected to manage the syndicate (originating house) and to underwrite the major amount of the issue. One bid price for the issue is made on behalf of the group, but the terms and features of the issue are set by you.

The distribution channels for a new security issue appear in Figure 18.2.

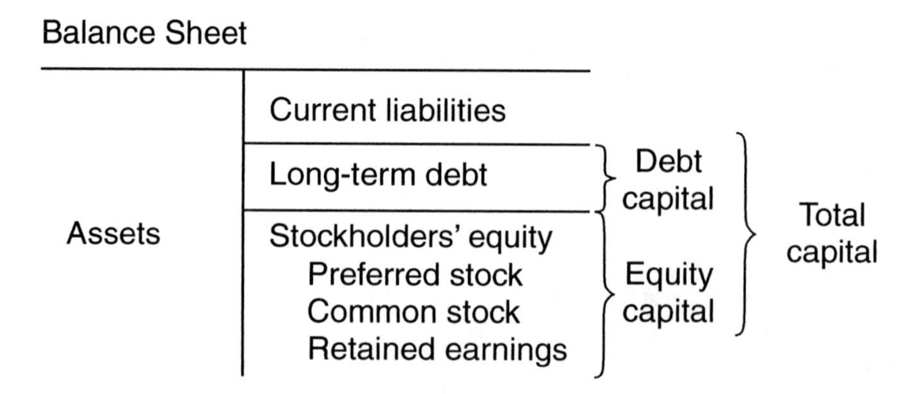

FIGURE 18.1 Financing assets through debt and equity.

In another approach to investment banking, the investment banker agrees to sell your securities on a best-efforts basis, or as an agent. Here, the investment banker does not act as underwriter but instead sells the stock and receives a sales commission. An investment banker may insist on this type of arrangement when he or she has reservations about the success of the security offering.

Besides investment bankers, there are firms that specialize in specific financial functions with respect to stock. A *dealer* buys securities and holds them in inventory for subsequent sale, expecting a profit on the spread. The *spread* is the price appreciation of the securities. A *broker* receives and forwards purchase orders for securities to the applicable stock exchange or over-the-counter market. The broker is compensated with a commission on each sale.

18.2 PUBLICLY AND PRIVATELY PLACED SECURITIES

Equity and debt securities may be issued either publicly or privately. Whether to issue securities publicly or privately will depend on the type and amount of financing your company requires.

In a public issuance, the shares are bought by the general public. In a private placement, you issue securities directly to either one or a few large investors. The large investors are financial institutions such as insurance companies, pension plans, and commercial banks.

Private placement has the following advantages relative to a public issuance:

- The flotation cost is less. Flotation cost is the expense of registering and selling the stock issue. Examples are brokerage commissions and underwriting fees. The flotation cost for common stock exceeds that for preferred stock. Flotation cost, expressed as a percentage of gross proceeds, is higher for smaller issues than for larger ones.
- It avoids SEC filing requirements.
- It avoids the disclosure of information to the public.
- There is less time involved in obtaining funds.

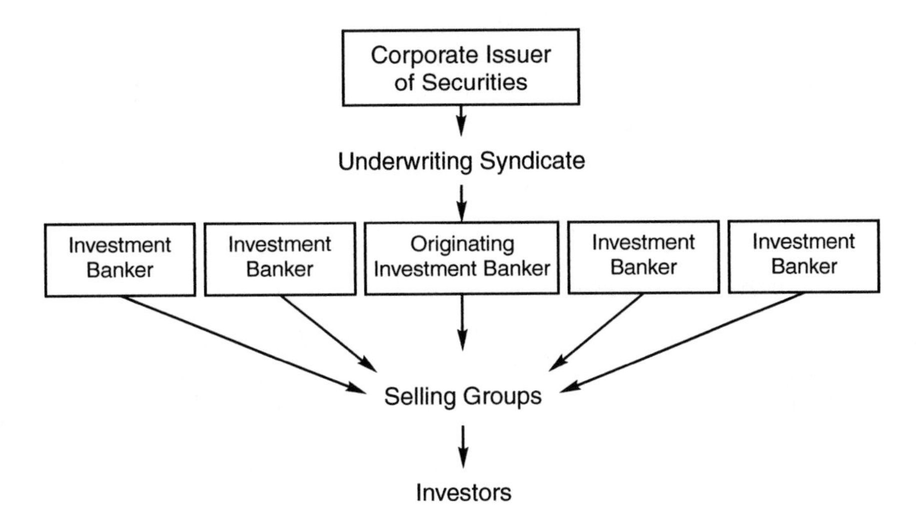

FIGURE 18.2 Distribution channels for a new security issue.

- It may not be practical to issue securities in the public market if you are so small that an investment banker would not find it profitable.
- Your credit rating may be low, and, as a result, investors may not be interested in purchasing securities when the money supply is low.

Private placement has the following drawbacks relative to a public issuance:

- It is more difficult to obtain significant amounts of money privately than publicly.
- Large investors typically use stringent credit standards, requiring the company to be in a strong financial position.
- Large institutional investors may watch the company's activities more closely than smaller investors in a public issuance.
- Large institutional investors are more capable of obtaining voting control of the company.

18.3 GOING PUBLIC — ABOUT AN INITIAL PUBLIC OFFERING (IPO)

Going public, often called an *initial public offering* (IPO), refers to selling formerly privately held shares to new investors on an organized exchange (e.g., New York Stock Exchange) or the over-the-counter market for the first time. For the individual company, going public marks a historic moment. It often is the springboard for greater growth and success. There are the advantages and disadvantages of raising capital through a public offering. The market for stocks of companies that are going public is called the *new issue market*.

The public sale of ownership interests can generate funds for business expansion, working capital, repayment of debt, diversification, acquisitions, marketing, and other uses. In addition, a successful IPO increases the visibility and appeal of your company, thereby escalating the demand and value for shares. Investors can benefit from an IPO not only because of the potential increase in market value for their stock, but also because publicly-held stock is more liquid and can be readily sold if the business appears to falter or if the investor needs quick cash. The availability of a public market for shares will also help determine the taxable values of the shares and assist in estate transfers.

18.3.1 How Does Going Public Work?

A company considering *going public* will typically work with an underwriter (an investment bank) who, either singly or as part of a group (syndicate), purchases the stock from the company (the issuer) at a discount from the public offering price. The underwriter or syndicate then sells the shares to the public through brokerage firms and other institutions. Creation of a syndicate pools the risk for underwriters and widens distribution channels for the new issue.

The underwriter advises the company on the marketability of, and demand for, its shares. Estimating the demand for the new stock issue is as much art as science. To estimate the demand, the underwriter and any dealers will collect *indications of interest* from their investors as to how many shares each would like to purchase.

Often, some IPOs may be oversubscribed, implying the demand for shares is larger than the number of shares to be issued. In the case of oversubscription, the underwriters determine who will get the shares at the public offering price. After the underwriter issues the shares, the shares begin trading on the open market. The laws of demand and supply take over. The price of IPO shares can increase or decrease substantially in a short time. For example, a company could have an IPO price of $15 per share, with the first trade in the open market executed at $80. Most of the players in the IPO market are fund managers from big institutions.

18.3.2 The Pros of Going Public

- Going public raises money — if it is common stock, it does not have to be repaid. The typical IPO raises $20 – 40 million, but offerings of $100 million are not unusual. This will vary widely by industry. Once public, companies can easily go back to the public market to raise more money. Typically, one third of all IPO issuers return to the public market within five years to issue a "seasoned equity offering." For example, in 1995 General Motors raised $1.14 billion by issuing new common stock. Since the shares sold were newly created, GM's issue was defined as a primary market offering, but since the firm was already publicly held, the offering was not an IPO. Firms generally prefer to obtain equity by retained earnings because of the flotation costs and market pressure involved with the sale of new common stock. However, if a firm needs more money than can be generated from retained earnings, a common stock offering may be necessary.
- As a company expands and becomes more valuable, its owners usually have most of their wealth tied up in the company. By selling some of their

stock in a public offering, they can reduce the riskiness of their personal portfolios through this diversification of their holdings.

- Management often experiences an increase in prestige and reputation. Public firms have higher profiles than private firms. This is important in industries where success requires suppliers and consumers to make long-term investments. For example, software requires training and no manager wants to buy software from a firm that may not be around for future improvements, upgrades, bug fixes, etc. The suppliers' and consumers' perception of company success is a self-fulfilling prophecy. However, public firms are usually bigger to begin with, and this may explain why public firms have a better image, on average. Going public will not increase its sales. The important question is if going public improves the company's stockholders' perception of success.
- Publicly traded stock can make a business more attractive to prospective and existing employees if stock option and other stock compensation plans are offered. Employee stock-based programs are worth more if transfer restrictions, such as those normally accompanying private company stock, are not placed on the stock.
- Mergers and acquisitions — many private companies do not just appear on the "radar screen" of potential acquirers. Being public makes it much easier for other firms to notice and analyze the company for synergy.
- The use of proceeds from the sale of the issue is generally unrestricted.
- Public companies can acquire other businesses with stock, without depleting cash reserves.
- Other financing alternatives may improve.

18.3.3 The Cons of Going Public

- Much jealously guarded information must be disclosed. The guarded items include management salaries, competitive position, transactions between the company and its management, and the identity of significant customers and suppliers. In addition to the required disclosure of results of operations and financial condition, public companies must be prepared to disclose information about the company, officers, directors, and certain shareholders. This information might include company sales and profits by product line, salaries and other compensation of officers and directors, data about major customers, competitive position, pending litigation, and related party transactions. By releasing the information it will become available to competitors, customers, employees, and the general public. The information is required in the initial registration statement and updated annually through annual reports, proxies, and other public disclosure documents. The company's IPO filings are with the Securities and Exchange Commission (on the Internet, **http://www.sec.gov**).
- In addition to the time and effort required to prepare for the filing and offering, a company must be prepared to incur the cost of going public. The principal costs include the underwriter's compensation, legal and

accounting fees, printing charges, and transfer agent and filing fees. A company expecting to go public with a high-quality offering should anticipate spending approximately $200,000, excluding underwriter's commissions. The magnitude of these costs usually make public offerings grossing less than $5 million impractical. Furthermore, principals must remember that there is no guarantee that the offering will be a success. With the exception of underwriter's compensation, the costs are incurred regardless of the outcome. The cost of going public does not stop with the initial offering. Other costs associated with being a public company are ongoing. Management must devote time and money to such new areas as shareholder relations, public relations, public disclosures, periodic filings with the SEC, and reviewing stock activity. All of this time, and the time of the personnel hired to handle these functions, would be spent on other management tasks in a privately held company. There are other out-of-pocket expenses. Shareholder meetings, annual and quarterly reports, public relations efforts and legal, accounting, and auditing fees must be paid. The total cost of these expenses vary from company to company, but in most cases they range from $50,000 to $150,000 annually.

- Corporate decision making becomes more cumbersome as the company attempts to move from a tightly controlled, entrepreneurially oriented company to a professionally managed one where ownership and management are divorced. Any decision, long-term or short-term, may be manifested promptly in the company's stock price. The company may worry constantly about improving quarterly earnings (and stock prices) instead of trying to take a longer perspective in developing its strategy.
- All IPO participants in the coalition are jointly liable for each others' actions. In practice, they were routinely sued for various omissions in the IPO prospectus when the public market valuation fell below the IPO offering price. In response, Congress passed *The Private Securities Litigation Reform Act of 1995*. This act protects the disclosure of firm projections and forces the suing shareholders to have substantial participation in the firm. Although nothing can eliminate lawsuits, this act reduces the likelihood of successful suits and therefore encourages settlement terms.
- Since the number of shares outstanding increases when the company goes public, greater earnings must be achieved to avoid reducing earnings per share.
- If the market price declines, many problems may result: management is usually personally blamed; the flexibility of issuing stock to make acquisitions may be hampered; if the decline occurs soon after the offering, litigation against everyone involved may take place; and other financing alternatives may evaporate.
- If the company is sitting on a gold mine, future earnings have to be shared with outsiders. After a typical IPO, about 40% of the company remains with insiders, but this can vary from 1 to 88%, with 20 to 60% being the comfortable norm.

- Outsiders can take control and even fire the entrepreneur. There is pressure on the managers to produce annual earnings gains, even when it may be in the shareholders' best long-term interests to adopt a strategy that reduces short-term earnings but raises them in the future years. These factors have led a number of newly public firms to go private in leveraged buyout deals where the managers borrow the money to buy out the non-management stockholders. The use of IPOs is limited primarily because:
  1. There is a very high cost and much complexity in complying with federal and state laws governing the sale of business securities (the cost for a small business can run from $50,000 to $500,000);
  2. Offering your business's ownership for public sale does little good unless your company has sufficient investor awareness and appeal to make the IPO worthwhile; and
  3. Management must be ready to handle the administrative and legal demands of widespread public ownership. Of course, an IPO also means a dilution of the existing shareholders' interests, and the possibility of takeovers or adjustments in management control are present.

18.3.4 How to Avoid the Drawbacks of Going Public

Here are some tips for avoiding the pitfalls of going public.

- Assemble the proper team. This involves selecting an underwriter, accountant, counsel, and perhaps some new directors.
- When choosing an underwriter, distribution capacity is important.
- An underwriter appropriate for one company or one industry may be inappropriate for another. In addition to technical ability, personalities and confidence should be considered.
- The selection of accountants and lawyers needs careful examination.
- The registration process is complex, coupled with absolute liability for the company for material misstatements or omissions — regardless of good faith or motive. It is important to remember that malpractice insurance in the securities field is the most expensive of any specialty. That carries a message. It is good to hire a "Big Five" or nationally prominent accounting firm because it enhances marketability and confidence. The use of a large accounting and legal firm may be viewed by the underwriter as insurance in the event of litigation.
- Securities laws are complicated. The sale of "securities" to the public is regulated by federal and state laws that have two primary objectives:
  1. Require businesses to disclose material information about the company to investors, and
  2. Prohibit misrepresentation and fraud in the sale of securities. Under federal law, a "security" is broadly defined and would include stocks, notes, bonds, evidence of indebtedness, and most ownership interests. The law defines a "public offering" of a security not by the number of investors to whom the stock is offered, but by the classification of

whether the investors are considered "sophisticated" or not. However, state law definitions of a "security" and of a "public offering" can vary from the federal law.

18.3.5 What Is the Process of Going Public?

A company that is thinking of going public should start acting like a public company as much as two years in advance. Several "to do's" include developing a business plan and preparing detailed financial results on a regular basis. Once a company decides to go public, it needs to choose its IPO team, consisting of a lead investment bank, accountant, lawyer, etc. The IPO process officially begins with an "all hands" meeting. This meeting usually takes place six to eight weeks before a company officially registers with the SEC. At this meeting, all members of the IPO team assign certain duties to each individual and plan a timetable for going public.

The most important and time-consuming task facing the IPO team is the development of the prospectus, which is a business document that basically serves as a brochure for the company. Since the SEC requires a "quiet period" on firms once they file an IPO until 25 days after a stock starts trading, the prospectus will have to do most of the talking and selling for the team. The prospectus includes all financial data of the company for the past five years. It also includes information on the management team and a description of a company's target market, competitors, and growth strategy. There is much more information in the prospectus, and the underwriting team goes to great efforts to make sure that it is all accurate.

The next step in the IPO process is the grueling worldwide tour, also known as the road show. The road show usually lasts a week, with company management going to a new city every day to meet with prospective investors and show off their business plan. The typical stops generally include the larger cities such as Los Angeles, San Francisco, Boston, and Chicago. If necessary, international destinations such as London and Hong Kong are included. The management team's performance on the road show is crucial. It helps determine the success of the IPO. The team has to impress institutional investors and influence them to make significant long-term investments.

Once the road show ends and the final prospectus is distributed to investors, management officials meet with their investment bank to decide on the final offering size and price. Investment banks generally suggest an appropriate price based on expected demand for the deal and other market conditions. The pricing of an IPO is a delicate balancing act. Investment firms have to be concerned with two different sets of customers: the company going public, which wants to raise as much money as possible, and the investors buying the shares of stock, who expect to see immediate gains for their money. If public interest appears to be slowing, it is common for the offering price and number of shares to decrease from expected ranges. Sometimes, a company even has to postpone an offering due to insufficient demand. If a deal is especially hot, the offering size and/or price can be raised from initial projections.

Once the offering price has been agreed to, an IPO is declared effective. This is generally done after a market closes, with the trading of the new stock beginning the next day. In the meantime, the chief underwriter works to confirm its buy orders. The chief underwriter is responsible for ensuring smooth trading during the first crucial days. This underwriter is legally allowed to support the price of a newly issued stock through buying shares in the market or selling them short. He can also impose penalty bids on brokers to discourage "flipping," which is when investors sell shares of an IPO soon after the stock starts trading. An IPO is not declared final until seven days after the firm's initial market appearance. On rare occasions, an IPO can be cancelled even after the stock begins trading.

18.3.6 Alternatives to Going Public

Many businesses can sell stock to insiders or to a small group of investors without being subject to securities laws; in effect, they can take advantage of alternatives to going public. However, it is not always clear where the exemptions end, so you should always consult a knowledgeable attorney before selling any stock in your company. The process of soliciting money from the public through the issuance and sale of securities requires a working knowledge of the state and federal registration statements concerning the securities to be sold, complex disclosure documents about the company with detailed information for potential investors, and financial statements. Employing professionals (attorney, accountant, and sometimes a stock underwriter) to assist in the process is a practical necessity.

While many small businesses sell interests in their companies that are "securities," as defined by federal or state laws, the transactions are often exempt from registration regulations because the offerings are sufficiently small in dollar amount, and they are restricted to a limited number and/or type of investor. These exempt offers of securities are called "limited private offerings" and they can avoid much of the cost and delay of a public offering. Unfortunately, to qualify for any of the exemptions, you must fit the criteria for both federal and state security laws. Limited private offerings can be either debt or equity instruments, or a hybrid of both. For instance, a convertible debt warrant would be a debt instrument that allows the holder to convert the debt into an equity interest at a certain time. These alternative offerings allow the business to tailor the amount of immediate equity (ownership and control) that it relinquishes, and the amount of debt (cash outflow) that it can safely assume. In this module, discussion of the use of limited private offerings is largely confined to equity financing.

Federal exemptions — At the federal level, the most popular exemption from registration requirements for small businesses is Rule 504, commonly known as "Regulation D." Under this provision, private companies that are selling less than $1 million worth of securities to any number of investors within a 12-month period are exempt from federal registration requirements. Solicitations of investors by a private business may be made through almost any means, including advertisements and seminars, and no specific disclosure requirements regarding the stock or the company are required. Most startups and smaller businesses would fall within this exemption.

Even if a securities offer is exempt from the registration requirements of federal or state law, the anti-fraud provisions of those laws may still apply. Therefore, you must take care to prevent misrepresentations or omissions in the offering that create an overly optimistic picture of the investment. The investor should be provided with sufficient information to make an informed decision regarding the investment.

Another exemption may be available to either private or publicly held companies that sell less than $5 million within a 12-month period, if the sales are made only to "accredited investors" and no more than 35 such investors are involved. *Accredited investors* include institutional investors (e.g., banks, brokers and dealers, insurance companies), company insiders (e.g., officers and directors), and wealthy investors ("wealthy" meaning they have more than $200,000 individual annual income or, individually or jointly with their spouse, have a net worth of over $1 million).

A lesser degree of exemption from regulation exists for a private or publicly held company that sells an unlimited issuance of securities to an unlimited number of accredited investors, or to no more than 35 nonaccredited but "sophisticated investors" (sophisticated investors have sufficient knowledge and experience so that they understand the risks of the sale, or the issuer reasonably believes the investors have these qualifications). Finally, an exemption exists for private offerings of stock that are sold only to persons living in the same state where the company is both incorporated and does significant business, although reliance upon this intrastate exemption is subject to continual policing because the securities must remain within the state.

State exemptions — Because each state has securities regulations, the local exemptions must be checked. Just because a sale may be exempt from federal registration does not mean state registration is not required. State securities laws are commonly referred to as "blue sky" laws because the regulations were originally enacted to prevent unscrupulous issuers from selling "speculative schemes that have no more basis than so many feet of blue sky." The state laws need not match the federal regulatory exemptions and even though a Uniform Securities Act exists for states to follow, that Act has not been adopted by each state nor is it consistently interpreted in those states which claim to follow it. The result is that consultation with a qualified professional is a practical necessity before soliciting investors for sales of securities. Forty-seven states currently have relaxed their securities regulations for small business by offering a Small Company Offering Registration (SCOR) procedure. Even if your business is not based in one of these states, you may still register and sell your securities in these states which have adopted SCOR. For a current list of eligible states, contact the North American Securities Administrators Association at 202-737-0900, or at **http://www.nasaa.org**.

18.4 VENTURE CAPITAL FINANCING

Venture capital firms supply funding from private sources for investing in select companies that have a high, rapid growth potential and a need for large amounts of capital. Venture capital (VC) firms speculate on certain high-risk businesses producing a very high rate of return in a very short time. The firms typically invest for periods of 3 to 7 years and expect at least a 20 to 40% annual return on their investment.

When dealing with venture capital firms, keep in mind that they are under great pressure to identify and exploit fast-growth opportunities before more conventional financing alternatives become available to the target companies. Venture capital firms have a reputation for negotiating tough financing terms and setting high demands on target companies. Three bottom-line suggestions:

- Make sure to read the fine print.
- Watch for delay maneuvers (they may be waiting for your financial position to weaken further).
- Guard your trade secrets and other proprietary information zealously.

Venture capital financing may not be available, nor a good choice of financing, for many businesses. Usually, venture capital firms favor existing businesses that have a minimal operating history of several years; financing of startups is limited to situations where the high risk is tempered by special circumstances, such as a company with extremely experienced management and a very marketable product or service. In 1995, venture capital firms invested in less than 2000 companies. The target companies often have revenues in excess of $2 million and a preexisting capital investment of at least $1 million.

VCs research target companies and markets more vigorously than conventional lenders, although the ultimate investment decision is often influenced by the market speculations of the particular venture capitalists. Due to the amount of money that venture capital firms spend in examining and researching businesses before they invest, they will usually want to invest at least a quarter of a million dollars to justify their costs.

Be wary of "shopping" innovative ideas to multiple venture capitalists or private investors. Use caution in revealing any information you consider proprietary. Even if you already have intellectual property protection (e.g., a patent, trademark, or copyright), you do not want to be forced to police your rights. Do your best to limit the details of your particular innovation and seek confidentiality arrangements for additional protection of any preexisting legal rights you may have.

The price of financing through venture capital firms is high. Ownership demands for an equity interest in 30 to 50% of the company are not uncommon even for established businesses, and a startup or higher risk venture could easily require transfer of a greater interest. Although the investing company will not typically get involved in the ongoing management of the company, it will usually want at least one seat on the target company's board of directors and involvement, for better or worse, in the major decisions affecting the direction of the company.

The ownership interest of the VC firm is usually a straight equity interest or an ownership option in the target company through either a convertible debt (where the debt holder has the option to convert the loan instrument into stock of the borrower) or a debt with warrants to a straight equity investment (where the warrant holder has the right to buy shares of common stock at a fixed price within a specified time period). An arrangement that eventually calls for an initial public offering is also possible. Despite the high costs of financing through venture capital companies, they offer tremendous potential for obtaining a very large amount of equity financing and they usually provide qualified business advice in addition to capital.

Venture capital firms are located nationwide, and a directory is available for $25 through the National Association of Venture Capital, 1655 N. Fort Meyer Dr., Arlington, VA 22209, (703-351-5269). In addition, other sources for venture capital can be found through bankers, insurance companies, and business associations.

18.5 TYPES OF LONG-TERM DEBT AND THEIR USEFULNESS

All companies incur debt; the amount of debt will largely depend on the company's available collateral. For long-term debt financing, sources include mortgages and bonds. Either instrument may be appropriate, depending on a company's circumstances. To determine which instrument will serve your company's needs best, you will need to know the characteristics, advantages, and disadvantages of these long-term debt sources.

18.5.1 MORTGAGES

Mortgages are notes payable that have real assets as collateral and request periodic payments. Mortgages can be issued to finance the purchase assets, construction of plant, and modernization of facilities. The bank requires that the value of the property exceed its mortgage. Most mortgage loans are made for 70 to 90% on the collateral. Mortgages may be obtained from a bank, a company, or another financial institution. It is easier to obtain loans for multiple-use real assets than for single-use real assets.

There are two types of mortgages: a *senior* mortgage, which has first claim on assets and earnings, and a *junior* mortgage, which has a subordinate lien.

A mortgage may have a closed-end provision that prevents you from issuing additional debt of the same priority against the same property. If the mortgage is open-ended, you can issue additional first-mortgage bonds against the property.

Mortgages have a number of advantages, including favorable interest rates, fewer financing restrictions, extended maturity dates for loan repayment, and relatively easy availability.

18.5.2 BONDS

Long-term debt principally takes the form of bonds payable and loans payable. A *bond* is a certificate indicating that you have borrowed a given sum of money and agree to repay it. A written agreement, called an *indenture*, describes the features of the bond issue (e.g., payment dates, call and conversion privileges [if any], and restrictions). The indenture is a contract between your company, the bondholder, and the trustee. The trustee makes sure that you meet the terms of the bond contract. In many instances, the trustee is the trust department of a commercial bank. Although the trustee is an agent for the bondholder, it is selected by your company prior to the issuance of the bonds. The indenture provides for certain restrictions on you such as a limitation on dividends and minimum working capital requirements. If a provision of the indenture is violated, the bonds are in default. The indenture may also have a negative pledge clause, which precludes the issuance of new debt taking priority over existing debt in the event of liquidation. The clause can apply to assets currently held as well as to assets that may be purchased in the future.

18.5.2.1 Computing Interest

Bonds usually come in $1,000 denominations. Many bonds have maturities of 10 to 30 years. The interest payment to the bondholder is called *nominal interest* (coupon interest, stated interest), which is the interest on the face value (maturity value, par) of the bond. Although the interest rate is stated on an annual basis, interest on a bond is typically paid semiannually. Interest expense is tax-deductible.

Example 18.1 — You issue a 15%, 10-year bond. The tax rate is 34%. The annual after-tax cost of the debt is 9.9% (15% × 66%).

Example 18.2 — You issue a $100,000, 8%, 10-year bond. The semiannual interest payment is $4,000 ($100,000 × 8% × 6/12). With a 34% tax rate, the after-tax semiannual interest is $2,640 ($4,000 × 66%).

A bond sold at face value is said to be sold at 100%. If a bond is sold below its face value (less than 100%), it is issued at a *discount*. If a bond is sold above face value (more than 100%), it is sold at a *premium*.

Why would your company's bond be sold at a discount or a premium? A bond may be sold at a discount when the interest rate on the bond is below the prevailing market interest rate for that type of security. It may also be issued at a discount if your company is risky, or there is a very long maturity period. A bond is issued at a premium when the opposite market conditions exist.

Example 18.3 — You issue a $100,000, 14%, 20-year bond at 94%. The maturity value of the bond is $100,000. The annual cash interest payment is $14,000 (14% × $100,000). The proceeds received for the issuance of the bond equal $94,000 ($100,000 × 94%). The amount of the discount is $6,000 ($100,000 – $94,000). The annual discount amortization is $300 ($6,000/20).

The yield on a bond is the effective (real) interest rate you incur. The two methods of computing yield are the simple yield and the yield-to-maturity.

The price of a bond depends on several factors such as maturity date, interest rate, and collateral. Bond prices and market interest rates are inversely related. For example, as market interest rates increase, the price of the existing bond falls because investors can invest in new bonds paying higher interest rates.

18.5.2.2 Types of Bonds

The various types of bonds your company may issue are as follows:

1. **Debentures** — Because debentures are unsecured (no collateral) debt, they can be issued only by large, financially strong companies with excellent credit ratings.
2. **Subordinated debentures** — The claims of the holders of these bonds are subordinated to those of senior creditors. Debt with a prior claim over the subordinated debentures is set forth in the bond indenture. Typically, in liquidation, subordinated debentures come after short-term debt.
3. **Mortgage bonds** — These are bonds secured by real assets. The first mortgage claim must be met before a distribution is made to a second

mortgage claim. There may be several mortgages for the same property (e.g., building).

4. **Collateral trust bonds** — The collateral for these bonds is your company's security investments in other companies (bonds or stocks), which are given to a trustee for safekeeping.
5. **Convertible bonds** — These may be converted to stock at a later date, based on a specified conversion ratio. The *conversion ratio* equals the par value of the convertible security divided by the conversion price. Convertible bonds are typically issued in the form of subordinated debentures. Convertible bonds are more marketable and are typically issued at a lower interest rate than regular bonds because they offer the *conversion right* to common stock. Of course, if bonds are converted to stock, debt repayment is not involved. A convertible bond is a quasi-equity security because its market value is tied to the value of the shares of stock into which the bond can be converted.
6. **Income bonds** — These bonds pay interest only if there is a profit. However, because it accumulates regardless of earnings, the interest, if bypassed, must be paid in a later year when adequate earnings exist.
7. **Guaranteed bonds** — These are debt issued by one party with payment guaranteed by another.
8. **Serial bonds** — A portion of these bonds comes due each year. At the time serial bonds are issued, a schedule shows the yields, interest rates, and prices for each maturity. The interest rate on the shorter maturities is lower than the interest rate on the longer maturities because less uncertainty exists regarding the future.
9. **Deep discount bonds** — These bonds have very low interest rates and thus are issued at substantial discounts. The return to the holder comes primarily from appreciation in price rather than from interest payments.
10. **Zero coupon bonds** — These bonds do not provide for interest. The return to the holder is in the form of appreciation in price.
11. **Variable-rate bonds** — The interest rates on the bonds are adjusted periodically to changes in money market conditions. These bonds are popular when there is uncertainty about future interest rates and inflation.

A summary of the characteristics and priority claims associated with bonds appears in Table 18.1.

18.5.2.3 Bond Ratings

Financial advisory services (e.g., Standard and Poor's, Moody's) rate publicly traded bonds according to risk in terms of the receipt of principal and interest. An inverse relationship exists between the quality of a bond issue and its yield. That is, low-quality bonds will have a higher yield than high-quality bonds. Hence, a risk-return tradeoff exists for the bondholder. Bond ratings are important because they influence market ability and the cost associated with the bond issue.

TABLE 18.1
Summary of Characteristics and Priority Claims of Bonds

Bond Type	Characteristics	Priority of Lender's Claim
Debentures	Available only to financially strong companies. Convertible bonds are typically debentures.	General creditor
Subordinated debentures	Comes after senior debt holders.	General creditor
Mortgage bonds	Collateral is real property or buildings.	Paid from the proceeds from the sale of the mortgaged assets. If any deficiency exists, general creditor status applies.
Collateral trust bonds	Secured by stock and/or bonds owned by the issuer. Collateral value is usually 30% more than bond value.	Paid from the proceeds of stock and/or bond that is collateralized. If there is a deficiency, general creditor status applies.
Income bonds	Interest is paid only if there is net income. Often issued when a company is in reorganization because of financial problems.	General creditor.
Deep-discount (and zero coupon) bonds	Issued at very low or no (zero) coupon rates. Issued at prices significantly below face value. Usually callable at par value.	Unsecured or secured status may apply depending on the features of the issue.
Variable-rate bonds	Coupon rate changes within limits based on changes in money or capital market rates. Appropriate when uncertainty exists regarding inflation and future interest rates. Because of the automatic adjustment to changing market conditions, the bonds sell near face value.	Unsecured or secured status may apply depending on the features of the issue.

18.5.3 The Advantages and Disadvantages of Debt Financing

You will need to determine whether or not your company should issue long-term debt. The advantages of issuing long-term debt include the following:

1. Interest is tax-deductible, while dividends are not.
2. Bondholders do not participate in superior earnings of your firm.
3. The debt is repaid in cheaper dollars during inflation.
4. There is no dilution of company control.
5. Financing flexibility can be achieved by including a call provision in the bond indenture. A call provision allows your company to pay the debt before the expiration date of the bond.
6. It may safeguard your company's future financial stability (e.g., in times of tight money markets when short-term loans are not available.)

The following disadvantages may apply to issuing long-term debt:

1. Interest charges must be met regardless of your company's earnings.
2. Debt must be repaid at maturity.
3. Higher debt implies greater financial risk, which may increase the cost of financing.
4. Indenture provisions may place stringent restrictions on your company.
5. Overcommitments may arise because of forecasting errors.

Should you buy a bond? To investors, bonds have the advantages of a fixed interest payment each year and of greater safety than equity securities because bondholders have a priority claim in the event of corporate bankruptcy. To investors, bonds have the disadvantages of not participating in incremental earnings and of no voting rights.

How does issuing debt stack up against equity securities to your company? The advantages of issuing debt rather than equity securities are as follows: interest is tax deductible, whereas dividends are not; during inflation, the payback will be in cheaper dollars; no dilution of voting control occurs; and flexibility in financing can be achieved by including a call provision in the bond indenture. The disadvantages of debt incurrence relative to issuing equity securities are that fixed interest charges and principal repayment must be met irrespective of your firm's cash flow position, and stringent indenture restrictions often exist.

The proper mixture of long-term debt to equity depends on company organization, credit availability, and the after-tax cost of financing. Where a high degree of debt already exists, you should take steps to minimize other corporate risks.

When should long-term debt be issued? Debt financing is more appropriate at these times:

1. The interest rate on debt is less than the rate of return earned on the money borrowed. By using other people's money (OPM), the after-tax profit of the company will increase. Stockholders have made an extra profit with no extra investment!
2. Stability in revenue and earnings exists so that the company will be able to meet interest and principal payments in both good and bad years. However, cyclical factors should not scare a company away from having any debt. The important thing is to accumulate no more interest and principal repayment obligations than can reasonably be satisfied in bad times as well as good.
3. There is a satisfactory profit margin so that earnings exist to meet debt obligations.
4. There is a good liquidity and cash flow position.
5. The debt/equity ratio is low so that the company can handle additional obligations.
6. Stock prices are currently depressed so that it does not pay to issue common stock at the present time.

7. Control considerations are a primary factor so that if common stock were issued, greater control might fall into the wrong hands.
8. Inflation is expected so that debt can be paid back in cheaper dollars.
9. Bond indenture restrictions are not burdensome.

Tip — If your company is experiencing financial difficulties, it may wish to refinance short-term debt on a long-term basis (e.g., by extending the maturity dates of existing loans). This may alleviate current liquidity and cash flow problems.

As the default risk of your company becomes higher, so will the interest rate to compensate for the greater risk.

Recommendation — When a high degree of debt (financial leverage) exists, try to reduce other risks (e.g., product risk) so that total corporate risk is controlled.

18.5.4 Bond Refunding

Bonds may be refunded before maturity through either the issuance of a serial bond or the exercise of a call privilege on a straight bond. The issuance of serial bonds allows you to refund the debt over the life of the issue. A call feature in a bond enables you to retire it before the expiration date. The call feature is included in many corporate bond issues.

When future interest rates are expected to drop, a call provision is recommended. Such a provision enables your firm to buy back the higher-interest bond and issue a lower-interest one. The timing for the refunding depends on expected future interest rates. A call price is typically set in excess of the face value of the bond. The resulting *call premium* equals the difference between the call price and the maturity value. Your company pays the premium to the bondholder in order to acquire the outstanding bonds prior to the maturity date. The call premium is usually equal to one year's interest if the bond is called in the first year, and it declines at a constant rate each year thereafter. Also involved in selling a new issue are flotation costs (e.g., brokerage commissions and printing costs).

A bond with a call provision typically will be issued at an interest rate higher than one without the call provision. The investor prefers not to face a situation where your company can buy back the bond at its option prior to maturity. The investor would obviously desire to hold on to a high-interest bond when prevailing interest rates are low.

Example 18.4 — A $100,000, 8%, 10-year bond is issued at 94%. The call price is 103%. Three years after the issue, the bond is called. The call premium is equal to

Call price	$103,000
Less: Face value of bond	(100,000)
Call premium	$ 3,000

The desirability of refunding a bond requires present value analysis, which was discussed in Chapter 12.

Example 18.5 — Your company has a $20 million, 10% bond issue outstanding that has 10 years to maturity. The call premium is 7% of face value. N 10-year bonds in the amount of $20 million can be issued at an 8% interest rate. Flotation costs of the new issue are $600,000.

Refunding of the original bond issue should occur as shown below.

Old interest payments ($20,000,000 × 0.10)	$2,000,000
Less: New interest payments ($20,000,000 × 0.08)	(1,600,000)
Annual savings	400,000
Call premium ($20,000,000 × 0.07)	1,400,000
Flotation cost	600,000
Total cost	$2,000,000

Year	Calculation	Present Value
0	–$2,000,000 × 1	–$2,000,000
1–10	$400,000 × 6.71[a]	2,684,000
	Net present value	$ 684,000

[a] 6.71 = T_4 (8%, 10 years) from Table 11.4.

Sinking fund requirements may exist in a bond issue. With a sinking fund, you put aside money to buy and retire part of a bond issue each year. Usually, there is a mandatory fixed amount that must be retired, but occasionally the retirement may relate to your company's sales or profit for the current year. If a sinking fund payment is not made, the bond issue may be in default.

18.6 EQUITY SECURITIES

We will now discuss equity financing consisting of preferred stock and common stock. The advantages and disadvantages of issuing preferred and common stock are addressed, along with the various circumstances in which one financing source is better suited than the other. Stock rights are also described.

18.6.1 PREFERRED STOCK

Preferred stock is a hybrid between bonds and common stock. Preferred stock comes after debt but before common stock in liquidation and in the distribution of earnings. Preferred stock may be issued when the cost of common stock is high. It is best to issue preferred stock when your company has excessive debt and an issue of common stock might result in control problems. Preferred stock is a more expensive way to raise capital than a bond issue because the dividend payment is not tax-deductible.

Preferred stock may be cumulative or noncumulative. *Cumulative* preferred stock means that if any prior year's dividend payments have been missed, they must be paid before dividends can be paid to common stockholders. If preferred dividends are in arrears for a long time, your company may find it difficult to resume its dividend payments to common stockholders. With *noncumulative* preferred stock, your company does not need to pay missed preferred dividends. Preferred stock dividends are limited to the rate specified, which is based on the total par value of the outstanding shares. Most preferred stock is cumulative.

Example 18.6 — As of December 31, 2005, your company had 6,000 shares of $15 par value, 14% cumulative preferred stock outstanding. Dividends were not paid in 2003 and 2004. Assuming that your company was profitable in 2005, the amount of the dividend to be distributed is:

Par value of stock = 6,000 shares × $15 = $90,000	
Dividends in arrears ($90,000 × 14% × 2 years)	$25,200
Current year dividend ($90,000 × 14%)	12,600
Total dividend	$37,800

Participating preferred stock means that if declared dividends exceed the amount typically given to preferred stockholders and common stockholders, the preferred and common stockholders will participate in the excess dividends. Unless stated otherwise, the distribution of the excess dividends will be based on the relative total par values. *Nonparticipating* preferred stock does not participate with common stock in excess dividends. Most preferred stock is nonparticipating.

Preferred stock may be callable, which means that you can purchase it back at a subsequent date at a specified call price. The call provision is advantageous when interest rates decline, since your company has the option of discontinuing payment of dividends at an excessive rate by buying back preferred stock issued when bond interest rates were high. Unlike bonds, preferred stock rarely has a maturity date. However, if preferred stock has a sinking fund associated with it, this, in effect, establishes a maturity date for repayment.

In bankruptcy, preferred stockholders are paid after creditors and before common stockholders. In such a case, preferred stockholders receive the par value of their shares, dividends in arrears, and the current year's dividend. Any asset balance then goes to the common stockholders.

The cost of preferred stock usually follows changes in interest rates. Hence, the cost of preferred stock will most likely be low when interest rates are low. When the cost of common stock is high, preferred stock issuance may be achieved at a lower cost.

The cost of preferred stock can be determined by dividing the dividend payment by the net proceeds received.

Example 18.7 — You issued preferred stock of $2 million. The flotation cost is 11% of gross proceeds. The dividend rate is 14%. The effective cost of the preferred stock is

$$\frac{\text{Dividend}}{\text{Net proceeds}} = \frac{0.14 \times \$2{,}000{,}000}{\$2{,}000{,}000 - (0.11 \times \$2{,}000{,}000)}$$

$$= \frac{\$280{,}000}{\$1{,}780{,}000} = 15.7\%$$

To your company, a preferred stock issue has the following advantages:

1. Preferred dividends do not have to be paid (important during financial distress), whereas interest on debt must be paid.
2. Preferred stockholders cannot force your company into bankruptcy.
3. Preferred stockholders do not share in unusually high profits because common stockholders are the real owners of the business.
4. If your company is growth oriented, it can generate better earnings for its original owners by issuing preferred stock with a fixed dividend rate than by issuing common stock.
5. Preferred stock issuance does not dilute the ownership interest of common stockholders in terms of earnings participation and voting rights.
6. Your company does not have to collateralize its assets as it may have to do if bonds are issued.
7. The debt to equity ratio is improved.

To your company, a preferred stock has the following disadvantages:

1. Preferred stock requires a higher yield than bonds because of greater risk.
2. Preferred dividends are not tax-deductible.
3. There are higher flotation costs than with bonds.

The advantages of preferred stock over bonds is that one can omit a dividend readily, no maturity date exists, and no sinking fund is required. Preferred stock also has a number of advantages over common stock. It avoids dilution of control and the equal participation in profits that are afforded to common stockholders. Disadvantages of preferred stock issuance compared to bonds are that it requires a higher yield than debt because it is riskier to the holder, and dividends are not tax deductible.

To an investor, a preferred stock offers several advantages. First, preferred stock usually provides a fixed dividend payment. Second, preferred stockholders' claims come before common stockholders' in the event of corporate bankruptcy. Third, preferred dividends are subject to an 80% dividend exclusion for *corporate* investors. For example, if a company holds preferred stock in another company and receives dividends of $10,000, only 20% (or $2,000) is taxable. On the other hand, interest income on bonds is fully taxable. Unfortunately, the *individual* investor does not qualify for the 80% dividend exclusion.

To an investor, preferred stock represents several disadvantages. For instance, the return is limited because of the fixed dividend rate. Also, there is greater price fluctuation with preferred stock than with bonds because of the lack of a maturity date. Lastly, preferred stockholders cannot require the company to pay dividends if the firm has insufficient earnings.

18.6.2 Common Stock Features

We will now discuss everything you should know about common stock. The owners of a corporation are called stockholders. They elect the board of directors, who in turn choose the company's officers. When the election takes place, management sends proxy statements, which ask stockholders to give management the right to vote their stock. Effective control of the corporation can exist with less than 50% common stock ownership since many stockholders do not bother to vote. Stockholders have limited liability in that they are not personally liable for the debts of the firm.

Common stock refers to the residual equity ownership in the business. Common stockholders have voting power, but their claims come after preferred stockholders in receiving dividends and in liquidation. Common stock does not involve fixed charges, maturity dates, or sinking fund requirements.

Authorized shares are the maximum shares that can be issued according to the corporate charter. *Issued shares* represent the number of authorized shares that have been sold by the company. *Outstanding shares* are the issued shares held by the investing public. *Treasury stock* is stock that has been reacquired by the firm. It is not retired but instead held for possible future resale, a stock option plan, use in purchasing another company, or the prevention of a takeover by an outside group. Outstanding shares are thus equal to the issued shares less the treasury shares. Dividends are based on the outstanding shares.

The *par value* of a stock is a stated amount of value per share specified in the corporate charter.

If your company is closely held, it only has a few stockholders, who keep full control. There is no requirement to publicly disclose corporate financial data. But a company with 500 or more stockholders must file an annual financial statement with the SEC.

A company may issue different classes of common stock. Class A is stock issued to the public and typically has no dividends. However, it usually has voting rights. Class B stock is typically kept by the company's organizers. Dividends are usually not paid on it until the company has generated adequate earnings. Voting rights are provided in order to maintain control.

The price of common stock moves in the opposite direction of market interest rates. For example, if market interest rates increase, stock prices fall because investors will transfer funds out of stock into higher yielding money market instruments and bank accounts. Further, higher interest rates make it costly for a company to borrow, resulting in lower profits and the subsequent decline in stock price.

Common stockholders have the following rights:

1. The right to receive dividends.
2. The right to receive assets upon the dissolution of the business.
3. The right to vote.
4. The preemptive right to buy new shares of common stock before their sale to the general public. In this way, current stockholders can maintain their proportionate percentage ownership in the company.
5. The receipt of a stock certificate that evidences ownership. The stock certificate may then be sold by the holder to another in the secondary security market.
6. The right to inspect the company's books.

A stockholder may make a significant profit if another firm wishes to take the company over. The stockholder typically receives notice of the offer directly from his or her broker or by an announcement in the financial pages of a newspaper (e.g., *The Wall Street Journal*.)

Disadvantages of common stock ownership are that common stockholders' claims come last in the event of corporate bankruptcy and that dividends may be bypassed.

A number of options exist for equity financing in the case of *small businesses*, including the following:

1. Venture capital (investor) groups who invest in typically high-risk ventures.
2. Issuances directly to institutional investors (e.g., insurance companies, banks).
3. Issuances to relatives or friends.
4. Issuances to major customers and suppliers.

A determination of the number of shares that must be issued to raise adequate funds to satisfy a capital budget may be needed.

Example 18.8 — Your company currently has 650,000 shares of common stock outstanding. The capital budget for the upcoming year is $1.8 million.

Assuming that new stock may be issued for $16 a share, the number of shares that must be issued to provide the necessary funds to meet the capital budget is:

$$\frac{\text{Funds needed}}{\text{Market price per share}} = \frac{\$1{,}800{,}000}{\$16} = 112{,}500 \text{ shares}$$

Example 18.9 — Your company wants to raise $3 million in its first public issue of common stock. After its issuance, the total market value of stock is expected to be $7 million. Currently, there are 140,000 outstanding shares that are closely held.

You want to compute the number of new shares that must be issued to raise the $3 million. The new shares will have 3/7ths ($3 million/$7 million) of the outstanding shares after the stock issuance. Thus, current stockholders will be holding 4/7ths of the shares.

$$140{,}000 \text{ shares} = 4/7 \text{ of the total shares}$$

$$\text{Total shares} = 245{,}000$$

$$\text{New shares} = 3/7 \times 245{,}000 = 105{,}000 \text{ shares}$$

After the stock issuance, the expected price per share is:

$$\text{Price per share} = \frac{\text{Market value}}{\text{Shares outstanding}} = \frac{\$7{,}000{,}000}{245{,}000} = \$28.57$$

When a company initially issues its common stock publicly, it is referred to as "going public." The estimated price per share to sell the securities is equal to:

$$\frac{\text{Anticipated market value of the company}}{\text{Total outstanding shares}}$$

For an established company, the market price per share can be determined using the *Gordon's growth model* (recall from Chapter 12). The model is:

$$P_o = \frac{D_1}{r - g}$$

or

$$\frac{\text{Expected dividend}}{\text{Cost of capital} - \text{Growth rate in dividends}}$$

Example 18.10 — Your company expected the dividend for the year to be \$10 a share. The cost of capital is 13%. The growth rate in dividends is expected to be constant at 8%. The price per share is:

$$\text{Price per share} = \frac{\text{Expected dividend}}{\text{Cost of capital} - \text{Growth rate in dividends}}$$

$$= \frac{\$10}{0.13 - 0.08} = \frac{\$10}{0.05} = \$200$$

Another approach to pricing the share of stock for an existing company is through the use of the price/earnings (P/E) ratio, which is equal to

$$\frac{\text{Market price per share}}{\text{Earnings per share}}$$

Example 18.11 — Your company's earnings per share are \$7. It is expected that the company's stock should sell at eight times its earnings. The market price per share is therefore:

$$P/E = \frac{\text{Market price per share}}{\text{Earnings per share}}$$

$$\text{Market price per share} = \text{P/E multiple} \times \text{Earnings per share}$$

$$= 8 \times \$7 = \$56$$

You may want to determine the market value of your company's stock. There are a number of ways to accomplish this.

Example 18.12 — Assuming an indefinite stream of future dividends of \$300,000 and a required return rate of 14%, the market value of the stock equals

$$\text{Market value} = \frac{\text{Expected dividends}}{\text{Rate of return}} = \frac{\$300{,}000}{0.14} = \$2{,}142{,}857$$

If there are 200,000 shares, the market price per share is

$$\text{Market value} = \frac{\$2{,}142{,}857}{200{,}000} = \$10.71$$

Example 18.13 — Your company is considering a public issue of its securities. The average price/earnings multiple in the industry is 15. The company's earnings are $400,000. There will be 100,000 shares outstanding after the issuance of the stock. The expected price per share is:

$$\text{Total market value} = \text{Net income} \times \text{Price/earnings multiple}$$

$$= \$400{,}000 \times 15 = \$6{,}000{,}000$$

$$\text{Price per share} = \frac{\text{Market value}}{\text{Shares}} = \frac{\$6{,}000{,}000}{100{,}000} = \$60$$

If your company has *significant* debt, it would be better off financing with an equity issue to lower overall financial risk.

Financing with common stock has the following advantages: (1) there is no requirement to pay dividends; (2) there is no repayment date; and (3) a common stock issue improves your company's credit rating relative to the issuance of debt.

Financing with common stock has several disadvantages. For example, dividends are not tax-deductible. Also, ownership interest is diluted. The additional voting rights could vote to take control away from the current ownership group. Further earnings and dividends are spread over more shares outstanding. Lastly, the flotation costs of a common stock issue are more than with preferred stock and debt financing. It is always cheaper to finance operations from internally generated funds because financing out of retained earnings involves no flotation costs.

A summary comparison of bonds and common stocks is presented in Table 18.2.

Stockholders are typically better off when a company cuts back on dividends instead of issuing common stock as a source of additional funds. When earnings are retained rather than new stock issued, the market price per share of existing stock will rise, as indicated by higher earnings per share. Also, a company typically earns a higher rate of return than stockholders so that by retaining funds market price of stock should appreciate. One caution, however: lower dividend payments may be looked at negatively in the market and may cause a reduction in the market price of stock due to psychological factors.

18.6.2.1 Stock Rights

Stock rights are options to buy securities at a specified price at a later date. The preemptive right provides that existing stockholders have the first option to buy additional shares. Exercising this right permits investors to maintain voting control and protects against dilution in ownership and earnings.

TABLE 18.2
Summary Comparison of Bonds and Common Stock

Bonds	Common Stock
Bondholders are creditors.	Stockholders are owners.
No voting rights exist.	Voting rights exist.
There is a maturity date.	There is no maturity date.
Bondholders have prior claims on profits and assets in bankruptcy.	Stockholders have residual claims.
Interest payments represent fixed charges.	Dividend payments do not continue fixed charges.
Interest payments are deductible on the tax return.	There is no tax deductibility for dividend payments.
The rate of return required by bondholders is typically lower than that by stockholders.	The required rate of return by stockholders is typically greater than that by bondholders.

18.7 HOW SHOULD YOU FINANCE?

Some companies obtain most of their funds from issuing stock and from earnings retained in the business. Other companies borrow as much as possible and raise additional money from stockholders only when they can no longer borrow. Most companies are somewhere in the middle. Companies use different mixes of financing, depending on their circumstances.

In formulating a financing strategy in terms of source and amount, you should consider the following:

1. The cost and risk of alternative financing strategies.
2. The future trend in market conditions and how they will impact future fund availability and interest rates. For example, if interest rates are expected to go up, you would be wise to finance with long-term debt at the currently lower interest rates.
3. The current debt-to-equity ratio. A very high ratio, for example, indicates financial risk, so additional funds should come from equity sources.
4. The maturity dates of present debt instruments. For example, you should avoid making all debt come due at the same time because in an economic downturn you may not have adequate funds to meet all the debt.
5. The restriction in loan agreements. For instance, a restriction may exist placing a cap on the allowable debt-to-equity ratio.
6. The type and amount of collateral required by long-term creditors.
7. The ability to change financing strategy to adjust to changing economic conditions.
8. The amount, nature, and stability of internally generated funds. If stability exists in earnings generation, the company is better able to meet debt obligations.
9. The adequacy of present lines of credit for current and future needs.

10. The inflation rate because, with debt, the repayment is in cheaper dollars.
11. The earning power and liquidity position of the firm. For example, a liquid company is better able to meet debt payments.
12. The tax rate. For example, a higher tax rate makes debt more attractive because there is a greater tax savings from interest expense.

Example 18.14 — Your company is considering issuing either debt or preferred stock to finance the purchase of a plant costing $1.3 million. The interest rate on the debt is 15%. The dividend rate on the preferred stock is 10%. The tax rate is 34%.

The annual interest payment on the debt is

$$15\% \times \$1{,}300{,}000 = \$195{,}000$$

The annual dividend on the preferred stock is

$$10\% \times \$1{,}300{,}000 = \$130{,}000$$

The required earnings before interest and taxes to meet the dividend payment is

$$\frac{\$130{,}000}{(1 - 0.34)} = \$196{,}970$$

If your company anticipates earning $196,970 without a problem, it should issue the preferred stock.

Example 18.15 — Your company has sales of $30 million a year. It needs $6 million in financing for capital expansion. The debt-to-equity ratio is 68%. Your company is in a risky industry, and net income is not stable. The common stock is selling at a high P/E ratio compared to the competition. Under consideration is the issuance of either common stock or a convertible bond.

Because your company is in a high-risk industry and has a high debt-to-equity ratio and unstable earnings, the issuance of common stock is recommended.

Example 18.16 — Your company is a mature one in its industry. There is limited ownership. The company has vacillating sales and earnings. Your firm's debt-to-equity ratio is 70%, relative to the industry standard of 55%. The after-tax rate of return is 16%. Since your company is a seasonal business, there are certain times during the year when its liquidity position is inadequate. Your company is unsure about the best way to finance.

Preferred stock is one possible means of financing. Debt financing is not recommended because of the already high debt-to-equity ratio, the fluctuation in profit, and the deficient liquidity posture. Because of the limited ownership, common stock financing may not be appropriate because this would dilute the ownership.

Example 18.17 — Your company wants to construct a plant that will take about 1 year to construct. The plant will be used to produce a new product line, for which your company expects a high demand. The new plant will materially increase corporate size. The following costs are expected:

Cost to build the plant	$800,000
Funds needed for contingencies	$100,000
Annual operating costs	$175,000

The asset, debt, and equity positions of your company are similar to industry standards. The market price of the company's stock is less than it should be, taking into account the future earning power of the new product line. What would be the appropriate means to finance the construction?

Since the market price of stock is less than it should be and considering the potential of the product line, convertible bonds and installment bank loans might be appropriate means of financing because interest expense is tax-deductible. Additionally, the issuance of convertible bonds might not require repayment, since the bonds are likely to be converted to common stock because of the company's profitability. Installment bank loans can be gradually paid off as the new product generates cash inflow. Funds needed for contingencies can be in the form of open bank lines of credit.

If the market price of the stock was not at a depressed level, financing through equity would be an alternative financing strategy.

Example 18.18 — Your company wants to acquire another business but has not determined the optimal means to refinance the acquisition. The current debt-to-equity position is within the industry guideline. In prior years, financing has been achieved through the issuance of short-term debt.

Profit has shown vacillation and, as a result, the market price of the stock has fluctuated. Currently, however, the market price of the stock is strong.

Your company's tax bracket is low.

The purchase should be financed through the issuance of equity securities for the following reasons:

1. The market price of stock is currently at a high level.
2. The issuance of long-term debt will cause greater instability in earnings because of the high fixed interest charges. Consequently, there will be more instability in stock price.
3. The issuance of debt will result in a higher debt-to-equity ratio relative to the industry norm. This will negatively impact the company's cost of capital and availability of financing.
4. Because it will take a long time to derive the funds needed for the purchase price, short-term debt should not be issued. If short-term debt is issued, the debt would have to be paid before the receipt of the return from the acquired business.

18.7.1 Working a Loan Online

For borrowing money, lenders look for security in your assets and cash flow in addition to your character, so if you can show a healthy balance sheet and a decent credit rating, you can borrow money easily on the Internet.

Among the most useful sites for business borrowers are:

- Loanwise.com (**http://www.loanwise.com**), operated by Net Earnings, Inc. of San Mateo, CA;
- America's Business Funding Directory (**http://www.businessfinance.com**), operated by BFS, Inc. of Irvine, TX;
- Business Finance Mart (**http://www.bizfinance.com**), operated by Capital Resource & Financial, Inc. of Tallahassee, FL.

Loanwise is a web-based loan broker representing a number of banks nationwide — 10 at last count — with more signing on monthly. You fill out an online application detailing your business operation and your need for the loan, along with some personal information. Once the website checks your personal and business credit ratings electronically, you can get an answer the same day for loans under $50,000. Loans for larger amounts require additional documentation, usually submitted by mail, so the approval process takes longer.

Given good credit, you may get offers from more than one lender, allowing you to choose the one with the best terms. Interest rates and other terms reflect the risk, of course, but they generally reflect those available from any business bank.

Loanwise's lenders include American Express, the Bank of Hawaii, Irwin Union Bank, Provident Bank, Countrywide Business Alliance, Union Bank of California, Compass Bank, PNC Bank, Equilease Financial Services, and a nonbank small-business lender called Business Loan Center, Inc.

You cannot apply online with America's Business Funding Directory. Instead, the site gathers your business and personal information, probes a database of hundreds of lending sources nationwide, and lists those whose lending criteria appear to be a good fit for you. This website alerts the lenders to your query, prompting them to make contact with you. It also gives you the names and phone numbers of contacts at the lenders so that you can initiate the process yourself.

The site also offers useful pages detailing the ins and outs of commercial finance, including equipment leasing, government funds, investment funds, real estate finance, venture capital, small business development councils, and the Service Corps of Retired Executives, a nonprofit organization of business mentors and counselors (**http://www.score.org**).

The Business Finance Mart offers an online application for all kinds of loans, including factoring, sale-leaseback deals, loans guaranteed by the Small Business Administration, equipment leasing, lines of credit, and start-up loans for entrepreneurs. The website also has a bulletin board on which you can post items on buying or selling a business or buying or selling inventory or equipment.

In addition to these, lenders of every stripe maintain web pages, including all of the big banks targeting business customers, plus commercial finance companies and other sources of business loans. Many allow you to submit online applications for business loans.

There is also no end of sites offering advice to the business borrower. Netscape, for example, offers one highly useful site

http://www.netscape.com/netcenter/smallbusiness/business/finance

with a button ("Getting financing for your business") that takes you to pages outlining:

- The financing options open to you at different points in your business life cycle;
- The advantages and disadvantages of debt vs. equity financing;
- Sources of debt and equity financing; and
- Sources of government-backed financing.

The web page of ZD, Inc., publisher of *PC Magazine*,

http://www.zdnet.com/smallbusiness/filters/biz_2000/finance

details the process of seeking an SBA-backed loan and links you to sites from which you may download loan forms. The web pages of *Entrepreneur* magazine (**http://www.entrepreneurmag.com**), Intuit, Inc. (**http://www.quicken.com**), and *Inc.* magazine (**http://www.inc.com**) are also useful.

18.7.2 Raising Equity and Venture Capital Online

It is not possible to do an equity deal by remote control on the Internet. However, the Internet can help you simplify the first step in your search for equity or venture capital by getting your business plan in front of potential investors. A number of good websites seek to help you make these connections, among them:

- The Adventure Capital Register (**http://www.adventurecapital.com**) — A good place to start if you're looking for capital to launch a new business venture, invention, or idea. You can get tips on how to find funding or for links to funding sources.
- The Venture Capital Resource Library (**http://www.vfinance.com**), where you can post a synopsis of your business idea along with basic information about your business.
- The National Venture Capital Association (**http://www.nvca.org**), which lists over 300 venture capital and private equity firms along with links to many of their websites, allowing you to gauge the firms' likes and dislikes.
- The National Financial Services Network (**http://www.nfsn.com**), which allows you to identify specific banks, investment banks, venture capital firms, insurers and other sources of finance by state.
- Commercial Finance Online (**http://www.cfol.com**), where you can search through a large database for financing sources likely to have an interest in your business and its capital needs.

- The Capital Network (**http://www.thecapitalnetwork.com**), where you can search a database of financing sources or, as an alternative, profile your capital needs and match them to potential sources.
- The Angel Capital Electronic Network — ACE-Net for short (**http://ace-net.sr.unh.edu/pub**) — A site partly sponsored by the U.S. Small Business Administration. It is an Internet securities listing service to help entrepreneurs and investors find each other. The site is hosted by the University of New Hampshire.
- "Money Hunt" (**http://moneyhunter.com**) — A public television program to help entrepreneurs turn their ideas into reality. The site provides business plan templates to write that all-important plan for investors, a directory of relevant contacts, and advice from experts.
- Garage.com (**http://www.garage.com**), which matches entrepreneurs and investors through a rigorous and detailed "vetting" process probing business plans, management expertise, and the like.

18.8 CONCLUSION

Your company may finance long-term with debt or equity (preferred stock and common stock) funds. Each has its own advantages and disadvantages. The facts of a situation have to be examined to determine which type is best for your company's circumstances.

Part V

Dissecting Financial Statement Information

19 Understanding Financial Statements

Managers should have a good understanding of the company in order to make an informed judgment on the financial position and operating performance of the entity. The balance sheet, the income statement, and the statement of cash flows are the primary documents analyzed to determine the company's financial condition. The balance sheets give the company's position in terms of its assets, liabilities, and equity or net worth, while the income statement gives the company's sources of revenue, expenses, and net income. The statement of cash flows allows you to analyze the company's sources and uses of cash. These financial statements are included in the annual report.

A business entity is an economic unit that enters into business transactions that must be recorded, summarized, and reported. Each business must have a separate set of accounting records and a separate set of financial statements. The financial statements are the means of conveying to management and to interested outsiders a concise picture of the value and profitability of the business for a given period of time. An examination of what can be gained from these statements, and wherein the pitfalls lie, is useful in setting up a program or strategy for planning and controlling profits.

19.1 THE INCOME STATEMENT AND BALANCE SHEET

The income statement measures operating performance for a specified time period (such as for the year ended December 31, 2001). The income statement shows the revenue, expenses, and net income (or loss) for a period of definition of each element as follows.

19.1.1 Revenue

Revenue arises from the sale of merchandise (as by retail business), or the performance of services for a customer or a client (as by a lawyer). Revenue from sales of merchandise or sales of services is often identified merely as *sales*. Other terms used to identify sources of revenue include professional fees, commission revenue, and fares earned. When revenue is earned it results in an increase in either cash or accounts receivable.

19.1.2 Expenses

Expenses result from performing those functions necessary to generate revenue. The amount of an expense is either equal to the cost of the goods sold, the value of the services received (e.g., salary expense), or the expenditures necessary for conducting business operations (e.g., rent expense) during the period.

19.1.3 Net Income (Loss)

Net income, also called *profits* or *earnings*, is the amount by which total revenue exceeds total expenses for the reporting period. It should be noted that revenue does not necessarily mean receipt of cash and expense does not automatically imply a cash payment. Note that net income and net cash flow (cash receipts less cash payments) are different. For example, taking out a bank loan will generate cash but this is not revenue since merchandise has not been sold nor have services been provided. Further, capital has not been altered because of the loan.

Example 19.1 — Joan Biehl is a self-employed consultant. For the month of May 2001, she earned income of $10,000 from services rendered. Her business expenses were: telephone $1,000, electricity $4,500, rent $2,000, secretarial salary $300, and office supplies used $400. Her income statement for the period is as follows:

Joan Biehl
INCOME STATEMENT
For the Month Ended May 31, 2001

Revenue from professional services		$10,000
Less: Operating expenses		
Telephone	$1,000	
Electricity	500	
Rent	2,000	
Secretarial salary	300	
Office supplies	400	
Total operating expenses		4,200
Net income		$ 5,800

Note that each revenue and expense item has its own account. This specifically enables one to better evaluate and control revenue and expense sources and to examine relationships among account categories. For instance, the ratio of telephone expenses to revenue is 10% ($1,000/$10,000). If in the previous month the relationship was 3%, Joan Biehl would, no doubt, attempt to determine the cause for this significant increase.

The balance sheet, on the other hand, portrays the financial position of the company at a particular point in time. It shows what is owed (assets), how much is owed (liabilities), and what is left (assets minus liabilities, known as stockholders' equity or net worth). With the balance sheet, you cut the point, freeze the action, and want to know about the company's financial position as of a certain date (like 12/31/2001, the end of the reporting year). It is a snapshot, while the income statement is a motion picture.

19.1.4 Assets

Assets are economic resources which are owned by an organization and are expected to benefit future operations. Assets may have definite physical form such as buildings, machinery, or supplies. On the other hand, some assets exist not in physical or tangible form, but in the form of valuable legal claims or rights, such as *accounts receivables* from customers and *notes receivables* from debtors.

Assets which will be converted into cash within 1 year are classified as *current.* Examples are cash, marketable securities, receivables, inventory, and prepaid expenses. Prepaid expenses include supplies on hand and advance payments of expenses such as insurance and property taxes.

Assets having a life exceeding 1 year are classified as *noncurrent.* Examples are long-term investments, equipment, and buildings. Equipment and buildings are often called *plant assets* or *fixed assets.*

19.1.5 Liabilities

Liabilities are debts owed to outsiders (creditors), and are frequently described on the balance sheet by titles that include the word "payable." The liability arising from the purchase of goods or services on credit (on time) is called an *accounts payable.* The form of the liability when money is borrowed is usually a *note payable*, a formal written promise to pay a certain amount of money, plus interest, at a definite future time. *Accounts payable*, as contrasted to a *note payable*, does not involve the issuance of a formal promise written to the creditor, and it does not require payment of interest. Other examples of liabilities include various accrued expenses.

Liabilities payable within 1 year are classified as *current*, such as accounts payable, notes payable, and taxes payable. Obligations payable in a period longer than 1 year, for example, bonds payable, are termed *long-term liabilities.*

19.1.6 Equity

Equity is a residual claim against the assets of the business after the total liabilities are deducted. Capital is the term applied to the owners' equity in the business. Other commonly used terms for capital are *owners' equity* and *net worth.* In a sole proprietorship, there is only one capital account since there is only one owner. In a partnership, a capital account exists for each owner. In a corporation, capital represents the stockholders' equity, which equals the capital stock issued plus the accumulated earnings of the business (called retained earnings). There are two types of capital stock — common stock and preferred stock. Common stock entitles its owners to voting rights, while preferred stock does not. Preferred stock entitles its owners to priority in the receipt of dividends and in repayment of capital in the event of corporate dissolution.

Example 19.2 — The equity of the owners of the business is quite similar to the equity commonly referred to with respect to home ownership. If you were to buy a house for \$150,000 by putting down 20%, i.e., \$30,000 of your own money and borrowing \$120,000 from a bank, you would say that your equity in the \$150,000 house was \$30,000.

The balance sheet may be prepared either in report form or account form. In the report form, assets, liabilities, and capital are listed vertically. In the account form,

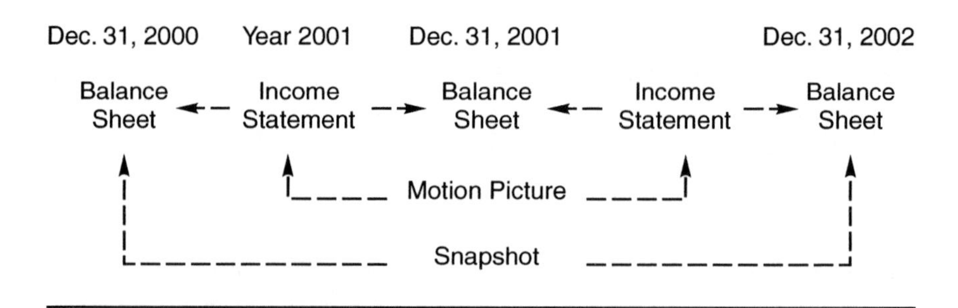

FIGURE 19.1 Balance sheet and income statement relationship.

assets are listed on the left side and liabilities and capital on the right side. From the examples given, it is evident that there is a tie-in between the income statement and the balance sheet. Biehl's net income of $5,800 (last item in her income statement from Example 19.1) is added to capital in her balance sheet in the above example. In effect, the income statement serves as the bridge between two consecutive balance sheets. Further, the net balance of the income statement accounts is used to adjust the capital account.

Figure 19.1 shows the relationship between the income statement and the balance sheet. In fact, the income statement serves as a bridge between the two consecutive balance sheets.

Example 19.3 — Report form.

Joan Biehl
Balance Sheet
May 31, 2001

ASSETS		
Cash		$10,000
Accounts receivable		20,000
Office supplies		10,500
Office equipment		30,000
Total assets		$71,000
LIABILITIES AND CAPITAL		
Liabilities		
Accounts payable		$30,000
Capital		
Balance, May 1, 2001	$35,600	
Net income for May	$5,800	
Less withdrawals	400	
Increase in capital	5,400	
Total capital		41,000
Total Liabilities and Capital		$71,000

Example 19.4 — Account form.

Joan Biehl
Balance Sheet
May 31, 2001

ASSETS		LIABILITIES AND CAPITAL		
Cash	$10,000	Liabilities		
Accounts receivable	20,000	Accounts payable		$30,000
Office supplies	10,500	Capital		
Office equipment	30,000	Balance, May 1, 2001	35,600	
		Net income for May	5,800	
		Less: Withdrawals	400	
		Increase in capital	5,400	
		Total capital		41,000
Total Assets	$71,000	Total Liabilities and Capital		$71,000

Note — Simply put, with the balance sheet you are asking "How wealthy or poor is the company?" while with the income statement you are asking "How did the company do last year?" and "Did it make money and then how much?" Neither one is good enough to tell you about the financial health of the company. For example, the fact that the company made a big profit does not necessarily mean it is wealthy, and vice versa. In order to get the total picture, you need both statements to complement each other.

Unfortunately, you still have problems. You would like to know more about the company's financial shape (such as the cash position of the company). However, neither the balance sheet nor the income statement provides the information of the cash flow during the period. The statement of cash flows provides this information, which will be discussed below.

19.2 THE STATEMENT OF CASH FLOWS

The *statement of cash flows* shows the sources and uses of cash, which are a basis for cash flow analysis for managers. The statement helps you to answer vital questions like "Where was money obtained?" and "Where was money put and for what purpose?" The following provides a list of more specific questions that can be answered by the statement of cash flows and cash flow analysis:

1. Is the company growing or just maintaining its competitive position?
2. Will the company be able to meet its financial obligations?
3. Where did the company obtain funds?
4. What use was made of net income?
5. How much of the required capital has been generated internally?
6. How was the expansion in plant and equipment financed?
7. Is the business expanding faster than it can generate funds?

8. Is the company's dividend policy in balance with its operating policy?
9. Is the company's cash position sound and what effect will it have on the market price of stock?

Cash is vital to the operation of every business. How management utilizes the flow of cash can determine a firm's success or failure. Financial managers must control their company's cash flow so that bills can be paid on time and extra dollars can be put into the purchase of inventory and new equipment or invested to generate additional earnings.

19.2.1 FASB Requirements

Management and external interested parties have always recognized the need for a cash flow statement. Therefore, in recognition of the fact that cash flow information is an integral part of both investment and credit decisions, the *Financial Accounting Standards Board* (FASB) has issued Statement No. 95, "Statement of Cash Flows." This pronouncement requires that enterprises include a statement of cash flows as part of the financial statements. A statement of cash flows reports the cash receipts, payments, and net change in cash on hand resulting from the *operating*, *investing*, and *financing* activities of an enterprise during a given period. The presentation reconciles beginning and ending cash balances.

19.2.2 Accrual Basis of Accounting

Under *Generally Accepted Accounting Principles* (GAAP), most companies use the accrual basis of accounting. This method requires that revenue be recorded when earned and that expenses be recorded when incurred. Revenue may include credit sales that have not yet been collected in cash and expenses incurred that may not have been paid in cash. Thus, under the accrual basis of accounting, net income will generally not indicate the net cash flow from operating activities. To arrive at net cash flow from operating activities, it is necessary to report revenues and expenses on a cash basis. This is accomplished by eliminating those transactions that did not result in a corresponding increase or decrease in cash on hand.

Example 19.5 — During 2001, the Eastern Electric Supply Corporation earned $2,100,000 in credit sales, of which $100,000 remained uncollected at the end of the calendar year. Cash that was actually collected by the corporation in 2001 can be calculated as follows:

Credit sales	$2,100,000
Less: Credit sales uncollected at year end	100,000
Actual cash collected	$2,000,000

A statement of cash flows focuses only on transactions involving the cash receipts and disbursements of a company. As previously stated, the statement of cash flows classifies cash receipts and cash payments into operating, investing, and financing activities.

19.2.3 Operating Activities

Operating activities include all transactions that are not investing or financing activities. They only relate to income statement items. Thus, cash received from the sale of goods or services, including the collection or sale of trade accounts and notes receivable from customers, interest received on loans, and dividend income are to be treated as cash from operating activities. Cash paid to acquire materials for the manufacture of goods for resale, rental payments to landlords, payments to employees as compensation, and interest paid to creditors are classified as cash outflows for operating activities.

19.2.4 Investing Activities

Investing activities include cash inflows from the sale of property, plant, and equipment used in the production of goods and services, debt instruments or equity of other entities, and the collection of principal on loans made to other enterprises. Cash outflows under this category may result from the purchase of plant and equipment and other productive assets, debt instruments or equity of other entities, and the making of loans to other enterprises.

19.2.5 Financing Activities

The financing activities of an enterprise involve the sale of a company's own preferred and common stock, bonds, mortgages, notes, and other short- or long-term borrowings. Cash outflows classified as financing activities include the repayment of short- and long-term debt, the reacquisition of treasury stock, and the payment of cash dividends.

Example 19.6 — The following information pertains to Liverpool Sugar Corporation during 2001.

1. The company had $1,004,000 in cash receipts from the sale of goods. Cash payments to acquire materials for the manufacture of goods totaled $469,000, its payments on accounts and notes payable amounted to $12,000, and it paid $136,000 in federal and state taxes.
2. The company sold all of its stock investment in Redondo Food Corporation, an unrelated entity, for $100,000. It then bought a new plant and equipment for $676,000.
3. In 2001, the company sold $300,000 of its 10%, 10-year bonds. It also issued another $50,000,000 in preferred stock in return for land and buildings. The company paid a cash dividend of $36,000.
4. The company had a $198,000 cash balance at the beginning of the year.

The statement cash flows for the company would be presented as follows:

Liverpool Sugar Corporation
Statement of Cash Flows
for the Year Ended December 31, 2001

Cash flows from operating activities:	
Cash received from customers	$ 1,004,000
Cash payment for acquisition of materials	(469,000)
Cash payment for interest and dividends	(12,000)
Cash payment for taxes	(136,000)
Net cash provided by operating activities	387,000
Cash flows from investing activities:	
Cash paid to purchase plant and equipment	(676,000)
Sale of long-term investment	(100,000)
Net cash provided by investing activities	(776,000)
Cash flows from financing activities:	
Sale of bonds	300,000
Cash paid for dividends	(36,000)
Net cash used in financing activities	264,000
Net decrease in cash and cash equivalents	(125,000)
Cash and cash equivalents at the beginning of the year	198,000
Cash and cash equivalents at the end of the year	$ 73,000

Note that the issuance of the preferred stock in exchange for the land and buildings is a noncash transaction that would be disclosed in supplementary form at the end of the statement of cash flows.

19.3 CONCLUSION

The traditional accounting statements — balance sheet, income statement, and the newly required statement of cash flows — have been and will continue to be the most important tools for both management and outsiders for use in gauging the financial condition of a business. As a later chapter will show, additional insights into the performance of the business can be gained by using financial ratio analysis.

20 Recording Financial Information and Accounting Conventions

Financial decisions are usually formulated on the basis of information generated by the accounting system of the firm. Proper interpretation of the data requires an understanding of the concepts and rules underlying such systems, the convention adopted in recording information, and the limitation inherent in the information presented.

The transactions of most businesses are numerous and complex, affecting many different items appearing on the financial statements. Therefore, a formal system of classification and recording is required for timely financial reporting and managerial needs. The aim of this chapter is to introduce the formal classification system of financial information commonly called *double-entry accounting*. By acquiring background information about this system, you will be able to more clearly understand the basic structure of the financial statements that were discussed in the previous chapter.

20.1 DOUBLE ENTRY AND THE ACCOUNTING EQUATION

Double-entry accounting is a system in which each business transaction affects and is recorded in two or more accounts with equal debits and credits.

20.1.1 THE ACCOUNTING EQUATION

An entity's financial position is reflected by the relationship between its assets and its liabilities and equity.

The accounting equation reflects these elements by expressing the equality of assets to creditors' claims and owners' equity as follows:

$$\text{Assets (A)} = \text{Liabilities (L)} + \text{Equity (E)}$$

The equation in effect says that a company's assets are subject to the rights of debt holders and owners.

The accounting equation is the basis for double-entry accounting, which means that each transaction has a dual effect. A transaction affects either both sides of the equation by the same amount or one side of the equation only, by increasing and decreasing it by identical amounts and thus netting zero.

Example 20.1 — Foster Architectural Company has assets of $700,000, obligations of $300,000, and owners' equity of $400,000. The accounting equation is:

Assets = Liabilities + Equity

\$700,000 = \$300,000 + \$400,000

If at the end of the reporting period, the firm derived net income of \$80,000, the accounting equation becomes

Assets = Liabilities + Equity

\$780,000 = \$300,000 + \$480,000

If \$60,000 was then used to pay creditors, the accounting equation becomes

Assets = Liabilities + Equity

\$720,000 = \$240,000 + \$480,000

In the next example, we will illustrate how the transactions of a business are recorded and what effect they have on the accounting equation.

Example 20.2 — Lloyd Thomas, a consultant, experienced the following events in the month of January 2001:

1. Started his consulting practice with a cash investment of \$10,000 and office equipment worth \$5,000;
2. Purchased office supplies of \$800 by paying cash;
3. Bought a word processor for \$500 on account from Smith Corona;
4. Paid \$400 in salary to his staff;
5. Received an electric bill for \$300;
6. Earned professional fees of \$20,000, of which \$12,000 was owed;
7. Paid \$300 to Smith Corona;
8. Withdrew \$100 from the firm for personal use; and
9. Received \$1,000 from one of the clients who owed him money.

The transactions will now be analyzed.

Transaction 1 — Thomas started his engineering consulting practice by investing \$10,000 in cash and \$5,000 in office equipment. The *assets* Cash and Office Equipment are increased, and the *equity* is also increased for the total investment of the owner.

Assets (A)		=	Liabilities (L)	+	Equity (E)
Cash	Office Equipment (OE)				L. Thomas Equity (E)
\$10,000	\$5,000				\$15,000

Transaction 2 — Acquired office supplies for cash, \$800.

The asset Office Supplies goes up by \$800 with a corresponding reduction in the asset Cash. This is an example of one asset being used to acquire another one.

A			=	L	+	E
Cash	OE	Office Supplies (OS)				
$10,000	$5,000					$15,000
–800		+$800				
$ 9,200	$5,000	$800				$15,000

Transaction 3 — Purchased a word processor on account, $500.

An *asset*, Office Equipment, is being acquired on credit, thereby creating a *liability* for future payment called Accounts Payable. Accounts Payable is defined as the amount owed to suppliers.

A			=	L	+	E
Cash	OE	OS		Accounts Payable (AP)		
$9,200	$5,000	$800				$15,000
	+500			+$500		
$9,200	$5,500	$800		$500		$15,000

Transaction 4 — Paid salary, $400.

Cash and equity are both being reduced because of the wage expense. Equity is reduced because expenses of the business decrease the equity of the owner.

A			=	L	+	E
Cash	OE	OS		AP		
$9,200	$5,500	$800		$500		$15,000
–400						–400
$8,800	$5,500	$800		$500		$14,600

Transaction 5 — Received an electric bill for $300 (not paid). Liabilities are increased by $300 since the firm *owes* the utility money for electricity supplied. Equity is reduced for the *expense*.

A			=	L	+	E
Cash	OE	OS		AP		
$8,800	$5,500	$800		$500		$14,600
				+300		–300
$8,800	$5,500	$800		$800		$14,300

Transaction 6 — Earned fees of $20,000, of which $12,000 was received in cash and $8,000 was owed by clients.

Cash goes up by $12,000 and the Accounts Receivable (amounts owed to the business from customers) is created. Professional fees earned are *revenue* to the business and hence increases the owner's equity. Thus, equity is increased by $20,000.

A				=	L	+	E
Cash	OE	OS	AR		AP		
$8,800	$5,500	$800			$800		$14,300
+12,000			+$8,000				+20,000
20,800	$5,500	$800	$8,000		$800		$34,000

Transaction 7 — Paid $300 to Smith Corona (in partial payment of the amount owed to them).

The payment lowers the asset Cash and reduces the liability Accounts Payable.

A				=	L	+	E
Cash	OE	OS	AR		AP		
$20,800	$5,500	$800	$8,000		$800		$34,300
–300					–300		
$20,500	$5,500	$800	$8,000		$500		$34,300

Transaction 8 — Withdrew $100 for personal use.

Cash is reduced and so is equity. The personal withdrawal is, in effect, a disinvestment in the business and hence reduces equity. It is not an expense in running the business.

A				=	L	+	E
Cash	OE	OS	AR		AP		
$20,500	$5,500	$800	$8,000		$500		$34,300
–100							–100
$20,400	$5,500	$800	$8,000		$500		$34,200

Transaction 9 — Received $1,000 from clients who owed him money.

This increases Cash and reduces Accounts Receivable since the client now owes the business less money. One asset is being substituted for another one.

A				=	L	+	E
Cash	OE	OS	AR		AP		
$20,400	$5,500	$800	$8,000		$500		$34,200
+1,000			–1,000				
$21,400	$5,500	$800	$7,000		500		$34,200

Transaction 10 — Worth (determined by an inventory count) of office supplies on hand at month's end, $600.

Since the worth of office supplies originally acquired was $800 and $600 is left on hand, the business used $200 in supplies. This reduces the asset Office Supplies and correspondingly reduces equity. The supplies used up represent office supplies expense.

A				=	L	+	E
Cash	OE	OS	AR		AP		
$21,400	$5,500	$800	$7,000		$500		$34,200
		–200					–200
$21,400	$5,500	$600	$7,000		$500		$34,000

SUMMARY OF TRANSACTIONS
January 2001

	A				=	L	+	E	
	Cash	OE	OS	AR		AP			
1.	$10,000	$5,000						$15,000	
2.	–800		+$800						
	9,200	5,000	800					15,000	
3.		+500				+$500			
	9,200	5,500	800			500		15,000	
4.	–400							–400	Wage Expense
	8,800	5,500	800			500		14,600	
5.						+300		–300	Utility Expense
	8,800	5,500	800			800		14,300	
6.	+12,000			$8,000				+20,000	Prof. Fee Revenue
	20,800	5,500	800	8,000		800		34,300	
7.	–300					–300			
	20,500	5,500	800	8,000		500		34,300	
8.	–100							–100	Drawing
	20,400	5,500	800	8,000		500		34,200	
9.	+ 1,000			–1,000					
	21,400	5,500	800	7,000		500		34,200	
10.			–200					–200	Office Supplies Expense
	$21,400	$5,500	$600	$7,000	=	$500		$34,000	
		$34,500			=		$34,500		

20.1.2 The Account

To prepare an equation Assets = Liabilities + Stockholders' Equity for each transaction would be extremely time consuming. Furthermore, information about a specific item (e.g., accounts receivable) would be lost through this process. Rather, there should be an *account* established for each type of item. At the end of the reporting period, the financial statements can then be prepared based upon the balances in these accounts.

The basic component of the formal accounting system is the *account*. A separate account exists for each item shown on the financial statements. Thus, balance sheet accounts consist of assets, liabilities, and equity. Income statement accounts are either expenses or revenue. The increases, decreases, and balances are shown for each account.

In other words, the purpose of the account is to provide a capsule summary of all transactions which have caused an increase or decrease and to reflect the account balance at any given point in time.

20.1.3 Ledger

All accounts are maintained in a book called the *ledger*. The ledger of a firm, for example, would be the group of accounts which summarize the financial operations of the company and is the basis for the preparation of the balance sheet and income statement. It is also useful for decision making since it provides the manager with the balance in a given account at a particular time.

20.1.4 A Chart of Accounts

The ledger is usually accompanied by a table of contents called a *chart of accounts*. The chart of accounts is a listing of the titles and numbers of all accounts in the ledger. Listed first are the balance sheet accounts — assets, liabilities, and stockholders' equity, in that order. The income statement accounts — revenue and expenses — follow. The account numbering system permits easy reference to accounts.

The account numbering system as it pertains to a typical company is as follows:

Series	Account Classification
001–199	Asset
200–299	Liabilities
300–399	Revenue
400–499	Expenses
500–599	Stockholders' Equities

Particular accounts can then be given unique, identifying account numbers within the series, as in the following examples from the Assets Series:

Account No.	Assets
001	Cash on hand
002	Marketable securities
003	Accounts receivable
004	Inventories
005	Investments
006	Land
007	Buildings
008	Equipment

Accounts may take many forms, but the simplest is called a *T-account*. The reason for this name is obvious, as shown below:

Account Title

debit (left side)	credit (right side)

Every account has three major parts:

1. A title, which is the name of the item recorded in the account;
2. A space for recording increases (monetary) in the amount of the item; and
3. A space for recording decreases (monetary) in the amount of the item.

The left and right sides of the account are called *debit* and *credit*, respectively (often abbreviated as "Dr" for debit and "Cr" for credit). Amounts entered on the left side of an account, regardless of the account title, are called debits to the account, and the account is said to be *debited*. Amounts entered on the right side of an account are called credits, and the account is said to be *credited*. You must note that the items debit and credit are *not* synonymous with the words increase and decrease. The system of debits and credits as related to increases and decreases in each of the six categories of accounts, assets, liabilities, revenue, expenses, and equity is explained later in the chapter.

To illustrate the account, we will look at the Cash account (within the Asset classification), where receipts of cash during a period of time have been listed vertically on the debit side and the cash payments for the same period have been listed similarly on the credit side of the account. A memorandum total of the cash receipts for the period to date, $55,000 in the illustration, may be noted below the last debit whenever the information is desired.

Cash

10,000	25,000
45,000	21,000
	7,500
55,000	53,500
1,500	

The total of the cash payments, $53,500 in the illustration, may be noted on the credit side in a similar manner. Subtraction of the smaller sum from the larger, $55,000 – $53,500, yields the amount of cash on hand, which is called the *balance* of the account. The cash account in the illustration has a balance of $1,500 (which may be inserted as shown), which identifies it as a *debit balance*.

20.1.5 The System of Debits and Credits

In this section we will briefly explain how accounts are increased or decreased through the use of debits and credits, the basic foundation of *double-entry* accounting where at least two entries, a debit and a credit, are made for each transaction.

The following guide shows how to increase or decrease accounts using debits and credits.

Asset	
+	–
Debit for increase	Credit for decrease

Liabilities	
–	+
Debit for decrease	Credit for increase

Revenue	
–	+
Debit for decrease	Credit for increase

Expenses	
–	+
Debit for increase	Credit for decrease

Equity	
+	–
Debit for decrease	Credit for increase

These same relationships are illustrated below:

Type of Account	Normal Balance	To Increase	To Decrease
Asset	Debit	Debit	Credit
Liability	Credit	Credit	Debit
Revenue	Credit	Credit	Debit
Expenses	Debit	Debit	Credit
Equity	Credit	Credit	Debit

The illustrated system of debits and credits is the standard method followed by persons keeping records on the double-entry system. The system of rules is analogous to a set of traffic rules whereby everyone (at least everyone in the U.S. and certain other nations) agrees to drive on the right side of the road. Obviously, the system would work if we reversed everything. However, you will see shortly that there is a very logical and unique system in the present structure.

20.1.6 The "How and Why" of Debits and Credits

Recall the fundamental accounting equation:

Assets (A) = Liabilities (L) + Equity (E)

In addition to this equation, there is another fundamental accounting concept or rule:

Debits must always equal Credits

This means that whenever a financial transaction is recorded in the accounting record, one account (or accounts) must be debited and another account (or accounts) must be credited to obtain an equal amount. It was noted earlier that all accounts have two sides, a debit side and a credit side. The purpose is to record increases on one side and decreases on the other.

20.1.7 Journals

For simplicity, the entries used in the previous section were made directly in the general ledger accounts. However, this process does not furnish the data required about a given transaction nor is listing of transactions in chronological order possible on T-accounts. These deficiencies are overcome through the use of a *journal*. The journal is the book of original entry in which transactions are entered on a daily basis in chronological order. This process is called *journalizing*.

The data are then transferred from the journal to the ledger by debiting and crediting the particular accounts involved. This process is called *posting*. The PR (Posting Reference) column is used for the ledger account number after the posting from the journal to the ledger takes place. This provides a cross reference between journal and ledger. There exist different types of journals that may be grouped into the categories of (1) general journals and (2) specialized journals. The latter is used when there are many repetitive transactions (e.g., sales or payroll).

20.2 CONCLUSION

We discussed some basic accounting conventions. A formal system of classification and recording — *double-entry bookkeeping* — was briefly covered. By acquiring background information about this system, you will be able to more clearly understand the basic structure of the financial statements that were discussed in the previous chapter.

21 Analyzing Financial Statements

Managers need to be able to analyze the company's financial statements in order to evaluate its financial health and operating performance. What has been the trend in profitability and return on investment? Will the business be able to pay its bills? How are the receivables and the inventory turning over? This chapter presents various financial statement analysis tools that are useful in evaluating the company's current and future financial conditions.

The analysis of financial statements reveals important information to management, creditors, and present and prospective investors. Financial statement analysis attempts to answer the following basic questions:

1. How well is the business doing?
2. What are its strengths?
3. What are its weaknesses?
4. How does it fare in the industry?
5. Is the business improving or deteriorating?

A complete set of financial statements, as explained in the previous chapter, will include the balance sheet, income statement, and statement of cash flows. The first two are vital in financial statement analysis. We will discuss the various financial statement analysis tools that you will use in evaluating the firm's present and future financial condition. These tools include horizontal, vertical, and ratio analysis, which give relative measures of the performance and financial condition of the company.

21.1 WHAT AND WHY OF FINANCIAL STATEMENT ANALYSIS

The analysis of financial statements means different things to different people. It is of interest to creditors, present and prospective investors, and the firm's own management.

A *creditor* is primarily interested in the firm's debt-paying ability. A short-term creditor, such as a vendor or supplier, is ultimately concerned with the firm's ability to pay its bills and therefore wants to be assured that the firm is liquid. A long-term creditor such as a bank or bondholder, on the other hand, is interested in the firm's ability to repay interest and principal on borrowed funds.

An *investor* is interested in the present and future level of return (earnings) and risk (liquidity, debt, and activity). You, as an investor, evaluate a firm's stock based on an examination of its financial statements. This evaluation considers overall financial health, economic and political conditions, industry factors, and future outlook of the company. The analysis attempts to ascertain whether the stock is overpriced, underpriced, or priced in proportion to its market value. A stock is

FIGURE 21.1
ALPHA, Inc.
Comparative Balance Sheet (in Thousands of Dollars)
December 31, 2002, 2001, 2000

				Increase or Decrease		% Increase or Decrease	
	2002	2001	2000	2002–2001	2001–2000	2002–2001	2001–2000
ASSETS							
Current assets							
Cash	$28	$36	$36	($8)	$0	–22.2%	0.0%
Marketable securities	22	15	7	7	8	46.7%	114.3%
Accounts receivable	21	16	10	5	6	31.3%	60.0%
Inventory	53	46	49	7	(3)	15.2%	–6.1%
Total current assets	124	113	102	11	11	9.7%	10.8%
Plant and equipment	103	91	83	12	8	13.2%	9.6%
Total assets	227	204	185	$23	$9	11.3%	10.3%
LIABILITIES							
Current liabilities	56	50	51	6	(1)	12.0%	–2.0%
Long-term debt	83	74	69	9	5	12.2%	7.2%
Total liabilities	139	124	120	$15	$4	12.1%	3.3%
STOCKHOLDERS' EQUITY							
Common stock, $10 par, 4,600 shares	46	46	46	0	0	0.0%	0.0%
Retained earnings	42	34	19	8	15	23.5%	78.9%
Total stockholders' equity	88	80	65	$8	$15	10.0%	23.1%
Total liabilities and stockholders' equity	$227	$204	$185	$23	$19	11.3%	10.3%

valuable to you only if you can predict the future financial performance of the business. Financial statement analysis gives you much of the data you will need to forecast earnings and dividends.

Management must relate the analysis to all of the questions raised by creditors and investors, since these interested parties must be satisfied for the firm to obtain capital as needed.

21.2 HORIZONTAL AND VERTICAL ANALYSIS

Comparison of two or more years' financial data is known as *horizontal analysis.* Horizontal analysis concentrates on the trend in the accounts over the years in dollar *and* percentage terms. It is typically presented in comparative financial statements (see ALPHA, Inc. financial data in Figures 21.1 and 21.2). In annual reports, comparative financial data are usually shown for five years.

Through horizontal analysis you can pinpoint areas of wide divergence requiring investigation. For example, in the income statement shown in Figure 21.3, the

FIGURE 21.2
ALPHA, Inc.
Comparative Income Statement (in Thousands of Dollars)
For the Years Ended December 31, 2002, 2001, 2000

				Increase or Decrease		% Increase or Decrease	
	2002	2001	2000	2002–2001	2001–2000	2002–2001	2001–2000
Sales	$98.3	$120	$56.6	($21.7)	$63.4	–18.1%	112.0%
Sales returns and allowances	18.0	10.0	4.0	8.0	6.0	80.0%	150.0%
Net sales	80.3	110.0	52.6	(29.7)	57.4	–27.0%	109.1%
Cost of goods sold	52.0	63.0	28.0	(11.0)	35.0	–17.5%	125.0%
Gross profit	28.3	47.0	24.6	(18.7)	22.4	–39.8%	91.1%
Operating expenses							
Selling expenses	12.0	13.0	11.0	(1.0)	2.0	–7.7%	18.2%
General expenses	5.0	8.0	3.0	(3.0)	5.0	–37.5%	166.7%
Total operating expenses	$17.0	$21.0	$14.0	($4.0)	$7.0	–19.0%	50.0%
Income from operations	$11.3	$26.0	$10.6	($14.7)	$15.4	–56.5%	145.3%
Nonoperating income	4.0	1.0	2.0	3.0	(1.0)	300.0%	–50.0%
Income before interest and taxes	15.3	27.0	12.6	(11.7)	14.4	–43.3%	114.3%
Interest expense	2.0	2.0	1.0	0.0	1.0	0.0%	100.0%
Income before taxes	13.3	25.0	11.6	(11.7)	13.4	–46.8%	115.5%
Income taxes (40%)	5.3	10.0	4.6	(4.7)	5.4	–46.8%	115.5%
Net income	$8.0	$15.0	$7.0	($7.0)	$8.0	–46.8%	115.5%

significant rise in sales returns taken with the reduction in sales for 2001–2002 should cause concern. You might compare these results with those of competitors.

It is essential to present both the dollar amount of change and the percentage of change, since the use of one without the other may result in erroneous conclusions. The interest expense from 2000–2001 went up by 100%, but this represented only $1,000 and may not need further investigation. In a similar vein, a large number change might cause a small percentage change and not be of any great importance. Key changes and trends can also be highlighted by the use of *common-size statements*. A common-size statement is one that shows the separate items in percentage terms. Preparation of common-size statements is known as *vertical analysis*. In vertical analysis, a material financial statement item is used as a base value, and all other accounts on the financial statement are compared to it. In the balance sheet, for example, total assets equal 100%. Each asset is stated as a percentage of total assets. Similarly, total liabilities and stockholders' equity is assigned 100% with a given liability or equity account stated as a percentage of the total liabilities and stockholders' equity, respectively. Figure 21.3 shows a common-size income statement based on the data provided in Figure 21.2.

Placing all assets in common-size form clearly shows the relative importance of the current assets as compared to the noncurrent assets. It also shows that significant

FIGURE 21.3
ALPHA, Inc.
Income Statement and Common Size Analysis (In Thousands of Dollars)
For the Years Ended December 31, 2002 and 2001

	2002 Amount	%	2001 Amount	%
Sales	$98.30	122.40%	$120.00	109.10%
Sales returns and allowances	18.00	22.40%	10.00	9.10%
Net sales	80.30	100.00%	110.00	100.00%
Cost of goods sold	52.00	64.80%	63.00	57.30%
Gross profit	28.30	35.20%	47.00	42.70%
Operating expenses				
Selling expenses	12.00	14.90%	13.00	11.80%
General expenses	5.00	6.20%	8.00	7.30%
Total operating expenses	$17.00	21.20%	$21.00	19.10%
Income from operations	$11.30	14.10%	$26.00	23.60%
Nonoperating income	4.00	5.00%	1.00	0.90%
Income before interest and taxes	15.30	19.10%	27.00	24.50%
Interest expense	2.00	2.50%	2.00	1.80%
Income before taxes	13.30	16.60%	25.00	22.70%
Income taxes (40%)	5.30	6.60%	10.00	9.10%
Net income	$8.00	9.90%	$15.00	13.60%

changes have taken place in the composition of the current assets over the last year. Notice, for example, that receivables have increased in relative importance and that cash has declined in relative importance. The deterioration in the cash position may be a result of inability to collect from customers.

For the income statement, 100% is assigned to net sales with all other revenue and expense accounts related to it. It is possible to see at a glance how each dollar of sales is distributed between the various costs, expenses, and profits. For example, notice from Figure 21.3 that 64.8 cents of every dollar of sales was needed to cover cost of goods sold in 2002, as compared to only 57.3 cents in the prior year; also notice that only 9.9 cents out of every dollar of sales remained for profits in 2002 — down from 13.6 cents in the prior year.

You should also compare the vertical percentages of the business to those of the competition and to the industry norms. Then you can determine how the company fares in the industry.

21.3 WORKING WITH FINANCIAL RATIOS

Horizontal and vertical analysis compares one figure to another within the same category. It is also vital to compare two figures applicable to different categories. This is accomplished by ratio analysis. In this section, you will learn how to calculate the various financial ratios and how to interpret them. The results of the ratio analysis will allow you:

1. To appraise the position of a business;
2. To identify trouble spots that need attention; and
3. To provide the basis for making projections and forecasts about the course of future operations.

Think of ratios as measures of the relative health or sickness of a business. Just as a doctor takes readings of a patient's temperature, blood pressure, heart rate, etc., you will take readings of a business's liquidity, profitability, leverage, efficiency in using assets, and market value. Where the doctor compares the readings to generally accepted guidelines such as a temperature of 98.6 degrees as normal, you make some comparisons to the norms.

To obtain useful conclusions from the ratios, you must make two comparisons:

Industry comparison — This will allow you to answer the question "How does a business fare in the industry?" You must compare the company's ratios to those of competing companies in the industry or with industry standards (averages). You can obtain industry norms from financial services such as Value Line, Dun and Bradstreet, and Standard and Poor's. Numerous online services such as AOL and MSN Money Central also provide these data.

Trend analysis — To see how the business is doing over time, you will compare a given ratio for one company over several years to see the direction of financial health or operational performance.

Financial ratios can be grouped into the following types: liquidity, asset utilization (activity), solvency (leverage and debt service), profitability, and market value.

21.3.1 Liquidity

Liquidity is the firm's ability to satisfy maturing short-term debt. Liquidity is crucial to carrying out the business, especially during periods of adversity. It relates to the short term, typically a period of one year or less. Poor liquidity might lead to higher cost of financing and inability to pay bills and dividends. The three basic measures of liquidity are: (a) net working capital, (b) the current ratio, and (c) the quick (acid-test) ratio.

Throughout our discussion, keep referring to Figures 21.1 and 22.2 to make sure you understand where the numbers come from.

Net working capital equals current assets minus current liabilities. Net working capital for 2002 is:

$$\text{Net working capital} = \text{Current assets} - \text{Current liabilities}$$

$$= \$124 - \$56$$

$$= \$68$$

In 2001, net working capital was \$63. The rise over the year is favorable.

The *current ratio* equals current assets divided by current liabilities. The ratio reflects the company's ability to satisfy current debt from current assets.

$$\text{Current ratio} = \frac{\text{Current assets}}{\text{Current liabilities}}$$

For 2002, the current ratio is:

$$\frac{\$124}{\$56} = 2.21$$

In 2001, the current ratio was 2.26. The ratio's decline over the year points to a slight reduction in liquidity.

A more stringent liquidity test can be found in the *quick (acid-test) ratio.* Inventory and prepaid expenses are excluded from the total of current assets, leaving only the more liquid (or quick) assets to be divided by current liabilities.

$$\text{Acid-test ratio} = \frac{\text{Cash + Marketable securities}}{\text{Current liabilities}}$$

The quick ratio for 2002 is:

$$\frac{\$28 + \$21 + \$22}{\$56} = 1.27$$

In 2001, the ratio was 1.34. A small reduction in the ratio over the period points to less liquidity.

The overall liquidity trend shows a slight deterioration as reflected in the lower current and quick ratios, although it is better than the industry norms (see Figure 21.4 for industry averages). But a mitigating factor is the increase in net working capital.

21.3.2 Asset Utilization

Asset utilization (activity, turnover) ratios reflect the way in which a company uses its assets to obtain revenue and profit. One example is how well receivables are turning into cash. The higher the ratio, the more efficiently the business manages its assets.

Accounts receivable ratios comprise the accounts receivable turnover and the average collection period.

The *accounts receivable turnover* provides the number of times accounts receivable are collected in the year. It is derived by dividing net credit sales by average accounts receivable.

You can calculate average accounts receivable by the average accounts receivable balance during a period.

$$\text{Accounts receivable turnover} = \frac{\text{Net credit sales}}{\text{Average accounts receivable}}$$

For 2002, the average accounts receivable is:

$$\frac{\$21 + \$16}{2} = \$18.5$$

The accounts receivable turnover for 2002 is:

$$\frac{\$80.3}{\$18.5} = 4.34$$

In 2001, the turnover was 8.46. There is a sharp reduction in the turnover rate pointing to a collection problem.

The *average collection period* is the length of time it takes to collect receivables. It represents the number of days receivables are held.

$$\text{Average collection period} = \frac{\text{365 days}}{\text{Accounts receivable turnover}}$$

In 2002, the collection period is:

$$\frac{365}{4.34} = 84.1 \text{ days}$$

It takes this firm about 84 days to convert receivables to cash. In 2001, the collection period was 43.1 days. The significant lengthening of the collection period may be a cause for some concern. The long collection period may be a result of the presence of many doubtful accounts, or it may be a result of poor credit management.

Inventory ratios are especially useful when a buildup in inventory exists. Inventory ties up cash. Holding large amounts of inventory can result in lost opportunities for profit as well as increased storage costs. Before you extend credit or lend money, you should examine the firm's *inventory turnover* and *average age of inventory.*

$$\text{Inventory turnover} = \frac{\text{Cost of goods sold}}{\text{Average inventory}}$$

The inventory turnover for 2002 is:

$$\frac{\$52}{\$49.5} = 1.05$$

For 2001, the turnover was 1.33.

$$\text{Average age of inventory} = \frac{365}{\text{Inventory turnover}}$$

In 2002, the average age is:

$$\frac{365}{1.05} = 347.6 \text{ days}$$

In the previous year, the average age was 274.4 days. The reduction in the turnover and increase in inventory age points to a longer holding of inventory. You should ask why the inventory is not selling as quickly.

The *operating cycle* is the number of days it takes to convert inventory and receivables to cash.

$$\text{Operating cycle} = \text{Average collection period} + \text{Average age of inventory}$$

In 2002, the operating cycle is:

$$84.1 \text{ days} + 347.6 \text{ days} = 431.7 \text{ days}$$

In the previous year, the operating cycle was 317.5 days. An unfavorable direction is indicated because additional funds are tied up in noncash assets. Cash is being collected more slowly.

By calculating the *total asset turnover,* you can find out whether the company is efficiently employing its total assets to obtain sales revenue. A low ratio may indicate too high an investment in assets in comparison to the sales revenue generated.

$$\text{Total asset turnover} = \frac{\text{Net sales}}{\text{Average total assets}}$$

In 2002, the ratio is:

$$\frac{\$80.3}{(\$204 + \$227)/2} = \frac{\$80.3}{\$215.5} = 0.37$$

In 2001, the ratio was 0.57 ($110/$194.5). There has been a sharp reduction in asset utilization.

ALPHA, Inc. has suffered a sharp deterioration in activity ratios, pointing to a need for improved credit and inventory management, although the 2002 ratios are not far out of line with the industry averages (see Figure 21.4). It appears that problems are inefficient collection and obsolescence of inventory.

21.3.3 Solvency (Leverage and Debt Service)

Solvency is the company's ability to satisfy long-term debt as it becomes due. You should be concerned about the long-term financial and operating structure of any firm in which you might be interested. Another important consideration is the size of debt in the firm's capital structure, which is referred to as *financial leverage*. (Capital structure is the mix of the *long term* sources of funds used by the firm).

Solvency also depends on earning power; in the long run a company will not satisfy its debts unless it earns profit. A leveraged capital structure subjects the company to fixed interest charges, which contributes to earnings instability. Excessive debt may also make it difficult for the firm to borrow funds at reasonable rates during tight money markets.

The *debt ratio* reveals the amount of money a company owes to its creditors. Excessive debt means greater risk to the investor. (Note that equity holders come after creditors in bankruptcy.) The debt ratio is:

$$\text{Debt ratio} = \frac{\text{Total liabilities}}{\text{Total assets}}$$

In 2002, the ratio is:

$$\frac{\$139}{\$227} = 0.61$$

The *debt-equity ratio* will show you if the firm has a great amount of debt in its capital structure. Large debts mean that the borrower has to pay significant periodic interest and principal. Also, a heavily indebted firm takes a greater risk of running out of cash in difficult times. The interpretation of this ratio depends on several variables, including the ratios of other firms in the industry, the degree of access to additional debt financing, and stability of operations.

$$\text{Debt-equity ratio} = \frac{\text{Total liabilities}}{\text{Stockholders' equity}}$$

In 2002, the ratio is:

$$\frac{\$139}{\$88} = 1.58$$

In the previous year, the ratio was 1.55. The trend is relatively static.

Times interest earned (interest coverage ratio) tells you how many times the firm's before-tax earnings would cover interest. It is a safety margin indicator in that it reflects how much of a reduction in earnings a company can tolerate.

$$\text{Times interest earned} = \frac{\text{Income before interest and taxes}}{\text{Interest expense}}$$

For 2002, the ratio is:

$$\frac{\$15.3}{\$2.0} = 7.65$$

In 2001, interest was covered 13.5 times. The reduction in coverage during the period is a bad sign. It means that less earnings are available to satisfy interest charges.

You must also note liabilities that have not yet been reported in the balance sheet by closely examining footnote disclosure. For example, you should find out about lawsuits, noncapitalized leases, and future guarantees.

As shown in Figure 21.4, the company's overall solvency is poor, relative to the industry averages, although it has remained fairly constant. There has been no significant change in its ability to satisfy long-term debt. Note that significantly less profit is available to cover interest payments.

21.3.4 PROFITABILITY

A company's ability to earn a good profit and return on investment is an indicator of its financial well-being and the efficiency with which it is managed. Poor earnings have detrimental effects on market price of stock and dividends. Total dollar net income has little meaning unless it is compared to the input in getting that profit.

The *gross profit margin* shows the percentage of each dollar remaining once the company has paid for goods acquired. A high margin reflects good earning potential.

$$\text{Gross profit margin} = \frac{\text{Gross profit}}{\text{Net sales}}$$

In 2002, the ratio is:

$$\frac{\$28.3}{\$80.3} = 0.35$$

The ratio was .43 in 2001. The reduction shows that the company now receives less profit on each dollar sales. Perhaps higher relative cost of merchandise sold is at fault.

Profit margin shows the earnings generated from revenue and is a key indicator of operating performance. It gives you an idea of the firm's pricing, cost structure, and production efficiency.

$$\text{Profit margin} = \frac{\text{Net income}}{\text{Net sales}}$$

The ratio in 2002 is:

$$\frac{\$8}{\$80.3} = 0.10$$

For the previous year, profit margin was .14. The decline in the ratio shows a downward trend in earning power. (Note that these percentages are available in the common size income statement as given in Figure 21.3).

Return on investment is a prime indicator because it allows you to evaluate the profit you will earn if you invest in the business. Two key ratios are the *return on total assets* and the *return on equity.*

The return on total assets shows whether management is efficient in using available resources to get profit.

$$\text{Return on total assets} = \frac{\text{Net income}}{\text{Average total assets}}$$

In 2002, the return is:

$$\frac{\$8}{(\$227 + \$204)/2} = 0.037$$

In 2001, the return was 0.077. There has been a deterioration in the productivity of assets in generating earnings.

The *return on equity* (ROE) reflects the rate of return earned on the stockholders' investment.

$$\text{Return on common equity} = \frac{\text{Net income available to stockholder}}{\text{Average stockholders' equity}}$$

The return in 2002 is:

$$\frac{\$8}{(\$88 + \$80)/2} = 0.095$$

In 2001, the return was 0.207. There has been a significant drop in return to the owners.

The overall profitability of the company has decreased considerably, causing a decline in both the return on assets and return on equity. Perhaps lower earnings were due in part to higher costs of short-term financing arising from the decline in liquidity and activity ratios. Moreover, as turnover rates in assets go down, profit will similarly decline because of a lack of sales and higher costs of carrying higher current asset balances. As indicated in Figure 21.4, industry comparisons reveal that the company is faring very poorly in the industry.

Table 21.1 shows industries with high return on equity (in excess of 20%).

21.3.5 Market Value

Market value ratios relate the company's stock price to its earnings (or book value) per share. Also included are dividend-related ratios.

TABLE 21.1
Industries with High Return on Equity (ROE) Rates (in Excess of 20%) 1998

Cars and trucks	62.4%
Personal care	31.0
Eating places	22.9
Food processing	24.8
Beverages	32.0
Business machines and services	20.6
Telephone	28.6

Source: *Corporate Scorecard*, by Business Week, McGraw-Hill, March 1999, pp. 75–91. Used with permission.

Earnings per share (EPS) is the ratio most widely watched by investors. EPS shows the net income per common share owned. You must reduce net income by the preferred dividends to obtain the net income available to common stockholders. Where preferred stock is not in the capital structure, you determine EPS by dividing net income by common shares outstanding. EPS is a gauge of corporate operating performance and of expected future dividends.

$$\text{EPS} = \frac{\text{Net income} - \text{Preferred dividend}}{\text{Common shares outstanding}}$$

EPS in 2002 is:

$$\frac{\$8{,}000}{4{,}600 \text{ shares}} = \$1.74$$

For 2001, EPS was $3.26. The sharp reduction over the year should cause alarm among investors. As you can see in Figure 21.4, the industry average EPS in 2002 is much higher than that of ALPHA, Inc. ($4.51 per share vs. $1.74 per share).

Table 21.2 provides a list of highly profitable companies in terms of EPS.

The *price-to-earnings* (P/E) ratio, also called *earnings multiple*, reflects the company's relationship to its stockholders. The P/E ratio represents the amount investors are willing to pay for each dollar of the firm's earnings. A high multiple (cost per dollar of earnings) is favored since it shows that investors view the firm positively. On the other hand, investors looking for value would prefer a relatively lower multiple (cost per dollar of earnings) as compared with companies of similar risk and return.

$$\text{Price/earnings ratio} = \frac{\text{Market price per share}}{\text{Earnings per share}}$$

TABLE 21.2
1998 Highly Profitable Companies (in Terms of EPS)

Ford	$17.76
U.S. Home	4.68
Alcoa	4.84
CIGNA	6.05
IBM	6.57
Washington Post	41.10

Source: *Corporate Scorecard*, by Business Week, McGraw-Hill, March 1999, pp. 75–91. Used with permission.

TABLE 21.3
P/E Ratios

Company	Industry	1998
Boeing	Aerospace	32
General Motors	Cars and Trucks	21
Goodyear	Tire and Rubber	13
Gap	Retailing	52
Intel	Semiconductor	37
Pfizer	Drugs and Research	88

Source: *Corporate Scorecard*, by Business Week, McGraw-Hill, March 1999, pp. 75–91. Used with permission.

Assume a market price per share of $12 on December 31, 2002, and $26 on December 31, 2001. The P/E ratios are:

$$2001\text{: } \frac{\$12}{\$1.74} = 6.9$$

$$2002\text{: } \frac{\$26}{\$3.26} = 7.98$$

From the lower P/E multiple, you can infer that the stock market now has a lower opinion of the business. However, some investors argue that a low P/E ratio can mean that the stock is undervalued. Nevertheless, the decline over the year in stock price was 54% ($14/$26), which should cause deep investor concern.

Table 21.3 shows price-to-earnings ratios of certain companies.

Book value per share equals the net assets available to common stockholders divided by shares outstanding. By comparing it to market price per share you can get another view of how investors feel about the business. The book value per share in 2002 is:

$$\text{Book value per share} = \frac{\text{Total stockholders' equity} - \text{Preferred stock}}{\text{Common shares outstanding}}$$

$$= \frac{\$88{,}000 - 0}{4{,}600} = \$19.13$$

In 2001, book value per share was $17.39.

The increased book value per share is a favorable sign, because it indicates that each share now has a higher book value. However, in 2002, market price is much less than book value, which means that the stock market does not value the security highly. In 2001, market price did exceed book value, but there is now some doubt in the minds of stockholders concerning the company. However, some analysts may argue that the stock is underpriced.

The *price/book value ratio* shows the market value of the company in comparison to its historical accounting value. A company with old assets may have a high ratio, whereas one with new assets may have a low ratio. Hence, you should note the changes in the ratio in an effort to appraise the corporate assets.

The ratio equals:

$$\text{Price/book value} = \frac{\text{Market price per share}}{\text{Book value per share}}$$

In 2002, the ratio is:

$$\frac{\$12}{\$19.13} = 0.63$$

In 2001, the ratio was 1.5. The significant drop in the ratio may indicate a lower opinion of the company in the eyes of investors. Market price of stock may have dropped because of a deterioration in liquidity, activity, and profitability ratios. The major indicators of a company's performance are intertwined (i.e., one affects the other) so that problems in one area may spill over into another. This appears to have happened to the company in our example.

Dividend ratios help you determine the current income from an investment. Two relevant ratios are:

$$\text{Dividend yield} = \frac{\text{Dividends per share}}{\text{Market price per share}}$$

$$\text{Dividend payout} = \frac{\text{Dividends per share}}{\text{Earnings per share}}$$

Table 21.4 shows the dividend payout ratios of some companies.

TABLE 21.4
Dividend Payout Ratios 1998

General Electric	1.2%
General Motors	2.3
Intel	0.1
Wal-Mart	0.4
Pfizer	0.5
Hewlett Packard	0.9

Source: MSN Money Central Investor (**http://investor.msn.com**), April 11, 1999. Used with permission.

There is no such thing as a "right" payout ratio. Stockholders look unfavorably upon reduced dividends because they are a sign of possible deteriorating financial health. However, companies with ample opportunities for growth at high rates of return on assets tend to have low payout ratios.

21.4 AN OVERALL EVALUATION — SUMMARY OF FINANCIAL RATIOS

As indicated in the chapter, a single ratio or a single group of ratios is not adequate for assessing all aspects of the firm's financial condition. Figure 21.4 summarizes the 2001 and 2002 ratios calculated in the previous sections, along with the industry average ratios for 2002. The figure also shows the formula used to calculate each ratio. The last three columns of the figure contain subjective assessments of ALPHA's financial condition, based on trend analysis and 2002 comparisons to the industry norms. (Five-year ratios are generally needed for trend analysis to be more meaningful, however.)

By appraising the trend in the company's ratios from 2001 to 2002, we see from the drop in the current and quick ratios that there has been a slight detraction in short-term liquidity, although they have been above the industry averages. But working capital has improved. A material deterioration in the activity ratios has occurred, indicating that improved credit and inventory policies are required. They are not terribly alarming, however, because these ratios are not way out of line with industry averages. Also, total utilization of assets, as indicated by the total asset turnover, shows a deteriorating trend.

Leverage (amount of debt) has been constant. However, there is less profit available to satisfy interest charges. ALPHA's profitability has deteriorated over the year. In 2002, it is consistently below the industry average in every measure of profitability. In consequence, the return on the owner's investment and the return on total assets have gone down. The earnings decrease may be partly due to the firm's high cost of short-term financing and partly due to operating inefficiency. The higher costs may be due to receivable and inventory difficulties that forced a decline in the liquidity and activity ratios. Furthermore, as receivables and inventory turn over less, profit will fall off from a lack of sales and the costs of carrying more in current asset balances.

FIGURE 21.4
ALPHA, Inc.
Summary of Financial Ratios — Trend and Industry Comparisons

Ratios	Definitions	2001	2002	Industry[a]	Evaluation[b] Ind.	Trend	Overall
LIQUIDITY							
Net working capital	Current assets – Current liabilities	63	68	56	good	good	good
Current ratio	Current assets/Current liabilities	2.26	2.21	2.05	OK	OK	OK
Quick (acid-test) ratio	(Cash + Marketable securities + Accounts receivable)/Current liabilities	1.34	1.27	1.11	OK	OK	OK
ASSET UTILIZATION							
Accounts receivable turnover	Net credit sales/Average accounts receivable	8.46	4.34	5.5	OK	poor	poor
Average collection period	365 days/Accounts receivable turnover	43.1 days	84.1 days	66.4 days	OK	poor	poor
Inventory turnover	Cost of goods sold/Average inventory	1.33	1.05	1.2	OK	poor	poor
Average age of inventory	365 days/Inventory turnover	274.4 days	347.6 days	N/A	N/A	poor	poor
Operating cycle	Average collection period + Average age of inventory	317.5 days	431.7 days	N/A	N/A	poor	poor
Total asset turnover	Net sales/Average total assets	0.57	0.37	0.44	OK	poor	poor
SOLVENCY							
Debt ratio	Total liabilities/Total assets	0.61	0.61	N/A	N/A	OK	OK
Debt-equity ratio	Total liabilities/Stockholders' equity	1.55	1.58	1.3	poor	poor	poor
Times interest earned	Income before interest and taxes/Interest expense	13.5	7.65	10	OK	poor	poor

FIGURE 21.4 (continued)
ALPHA, Inc.
Summary of Financial Ratios — Trend and Industry Comparisons

Ratios	Definitions	2001	2002	Industry[a]	Evaluation[b]		
					Ind.	Trend	Overall
PROFITABILITY							
Gross profit margin	Gross profit/Net sales	0.43	0.35	0.48	poor	poor	poor
Profit margin	Net income/Net sales	0.14	0.1	0.15	poor	poor	poor
Return on total assets	Net income/Average total assets	0.077	0.037	0.1	poor	poor	poor
Return on equity (ROE)	Earnings available to common stockholders/Average stockholders' equity	0.207	0.095	0.27	poor	poor	poor
MARKET VALUE							
Earnings per share (EPS)	(Net income – Preferred dividend)/Common shares outstanding	3.26	1.74	4.51	poor	poor	poor
Price/earnings (P/E) ratio	Market price per share/EPS	7.98	6.9	7.12	OK	poor	poor
Book value per share	(Total stockholders' equity – Preferred stock)/Common shares outstanding	17.39	19.13	N/A	N/A	good	good
Price/book value ratio	Market price per share/Book value per share	1.5	0.63	N/A	N/A	poor	poor
Dividend yield	Dividends per share/Market price per share						
Dividend payout	Dividends per share/EPS						

[a] Obtained from sources not included in this chapter.

[b] Represents subjective evaluation.

The firm's market value, as measured by the price-to-earnings (P/E) ratio, is respectable as compared with the industry — but it shows a declining trend.

In summary, it appears that the company is doing satisfactorily in the industry in many categories. The 2001–2002 period, however, seems to indicate that the company is heading for financial trouble in terms of earnings, activity, and short-term liquidity. The business needs to concentrate on increasing operating efficiency and asset utilization.

21.5 CONCLUSION

Financial statement analysis is an attempt to work with reported financial figures in order to determine a company's financial strengths and weaknesses. Most analysts favor certain ratios and ignore others. Each ratio should be compared to industry norms and analyzed in light of past trends. Financial analysis also calls for an awareness of the impact of inflation and deflation on reported income. Management must also recognize that alternate methods of financial reporting may allow firms with equal performance to report different results.

Index

A

B

D

E

F

G

H

I

J

K

L

M

Q

R

S

T

U

V

W

Y

Z